Corn & Capitalism

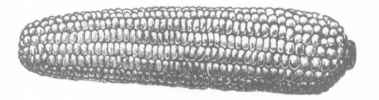

A book in the series
Latin America in Translation /
en Traducción /
em Tradução

Sponsored by the
Consortium in Latin American Studies at the
University of North Carolina at Chapel Hill
and Duke University

Corn & Capitalism

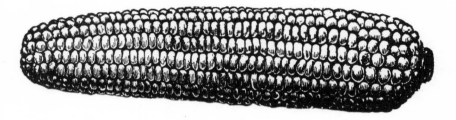

HOW A BOTANICAL BASTARD GREW

TO GLOBAL DOMINANCE

ARTURO WARMAN

TRANSLATED BY NANCY L. WESTRATE

The University of North Carolina Press

Chapel Hill and London

Originally published in Spanish with the title
La historia de un bastardo: maíz y capitalismo, © 1988, 1995,
Fondo de Cultura Económica, S.A. de C.V.
Manufactured in the United States of America
Set in Minion and Meta types by Keystone Typesetting, Inc.

Translation of the books in the series
Latin America in Translation / en Traducción / em Tradução,
a collaboration between the Consortium in Latin American Studies at the
University of North Carolina at Chapel Hill and Duke University and
the university presses of the University of North Carolina and Duke,
is supported by a grant from the Andrew W. Mellon Foundation.

The paper in this book meets the guidelines for permanence and
durability of the Committee on Production Guidelines for Book Longevity
of the Council on Library Resources.

Library of Congress Cataloging-in-Publication Data
Warman, Arturo.
[Historia de un bastardo]
Corn and capitalism : how a botanical bastard grew to global dominance /
Arturo Warman ; translated by Nancy L. Westrate.
p. cm. — (Latin America in translation/en traducción/em tradução)
Includes bibliographical references and index (p.   ).
ISBN 0-8078-2766-5 (cloth: alk. paper)
ISBN 0-8078-5437-9 (pbk.: alk. paper)
1. Corn—History. I. Title. II. Series.
SB191.M2 W34 2003
633.1′5′09—dc21
2002010956

cloth   07 06 05 04 03   5 4 3 2 1
paper   07 06 05 04 03   5 4 3 2 1

CONTENTS

TRANSLATOR'S PREFACE

Arturo Warman is familiar to U.S. audiences for his vivid portrayal of the peasants of Morelos in *We Come to Object*. That work, well known and highly regarded in the anthropological community, is in many ways the archetypal anthropological case study, the "small reality" Warman speaks of in his preface to the present volume. That case study is also typical in its mission to provide a voice for those routinely unable to express their outrage, demands, and ambitions. While *Corn and Capitalism* may appear at first to be something of a departure from more typical anthropological work, in many ways it provides exactly this same sort of outlet. This time, however, that voice belongs to a plant, corn, pivotal to the lives of countless people the world over. Like so many anthropological subjects, corn has alternately suffered, thrived under, and resisted the pressures of modernization, development, and the world marketplace.

Corn's place in the world today is the result of a number of complex historical interactions. This translation of Warman's work is the outcome of a similarly complex process. It is not a literal translation. Neither is it a paraphrase of the original Spanish edition. Rather, it is an English version filtered through differences of language and culture and the passing of years since the original edition. The translation is intended for a U.S. audience. This had led to certain excisions of material already part of our collective historical memory. Removing cultural stumbling blocks improved the book's readability in some cases. Certain accommodations were also demanded by the changes wrought over the years since Professor Warman first conducted the research for the book, which appeared in Spanish editions in 1988 and 1995. Thus the present translation updates certain time references and makes allowances for subsequent geopolitical changes. While this is not a comprehensively revised edition, material obviously no longer suitable has been gently excised in consultation with the author. Some theoretical language, including outmoded jargon, has been reworked or rephrased in order to maintain the underlying points that are still pertinent. Professor Warman intended the original text's Marxist jargon to be ambiguous, and these word choices have been carefully reviewed under his supervision. I have made other editorial changes at my own discretion, bowing to prevailing scholarly standards in this country. Some sections of the book remain generally faithful to the original, their point fully valid up into the early 1980s. Ultimately, I have been guided by what sounds best in English. Wherever possible I have tried to maintain the color and imagery Professor Warman employed throughout the original Spanish text.

I was raised in a small farming community in rural southwestern lower Michigan. My father worked for the United States Department of Agriculture as a soil conservationist and also farmed corn along with soy beans, wheat, and sorghum. With this background I found it eye-opening to learn about the imagery, status, and debates over corn that Warman so skillfully traces in these pages. As a native of the Midwest, I had never shared any of the taboos or stigmas associated with corn elsewhere in the world. I still am tremendously saddened every time I hear of yet another family farm I knew in my youth succumbing to the economic pressures of large-scale commercial agriculture. While my extended family still owns and operates those farms started by my father, that experience will probably end with this generation, as my siblings and our children move on to other endeavors far removed from the hardships and joys of shoulder-high corn rustling in a gentle afternoon breeze.

Many individuals have contributed to the accuracy and authenticity of this translation. I especially would like to thank my father-in-law, Jorge De Luca, for meticulously reviewing and commenting on the many versions of the manuscript. My sincerest thanks go out to the entire staff of Duke University's Perkins Library. I am especially grateful to the staffs of the Reference Department, Current Periodicals Department, and Documents Department. They were outstanding in their swift, professional, and extraordinarily competent responses to my seemingly unending queries. Of course, I would like to thank my family for their patience and the many accommodations they made as I worked to translate this book into the form it deserved. Last, it has been a tremendous experience to work with Professor Warman, exchanging ideas and perspectives on corn and culture.

PREFACE

There is a long story behind this book. I first began to methodically compile material on corn in the late 1970s. At that time, my intentions were quite open-ended and my interest in corn was more personal than programmatic. My original interest in corn was older still. Since I was a city dweller and, worse still, from Mexico City, corn was something I took for granted, something ubiquitous and constant, like the air we breathed or the water we drank. I discovered something in the rural countryside, if what millions already knew really could be deemed discovery: that peasants had created corn on a daily basis. They created corn by virtue of their hard work, their knowledge, and their respect and veneration of nature. Corn was a product of their passion, of their lives that revolved around that plant, and of their stubborn persistence. Their teachings made this book possible.

Shortly thereafter I discovered something more: that this same plant was a human invention, that nature could not propagate it without the participation of men, or more accurately of women, according to what archaeologists tell us. Bit by bit, it became clear that the story behind corn was far from the self-evident case I had originally thought it to be, and I began to delve into the mystery surrounding that plant. About twenty-five years ago, my curiosity, which had left me with a disordered array of papers and photocopies, became a methodical calling. I began to dig, to gather whatever information on corn I could to satisfy the hunger for pure knowledge. Perhaps this was my attempt to perform some sort of penance for my ignorance and urban arrogance, for paying so little attention to something that was at the center of the lives of millions of compatriots.

The accumulation of information inevitably became a sort of avarice. I intended to write a social history of corn in Mexico, that is to say, a history of the knowledge and the work needed to produce the material sustenance of an entire nation. That intention remains, and if life permits and I am afforded the time and opportunity, another book will follow this one. I will dedicate the sequel to analyzing Mexico's long social struggle for self-sufficiency and dietary self-determination. The very history of Mexico revolves around the struggle over corn. A research project that Carlos Montañez and I conducted on corn cultivation and corn producers in modern Mexico partially served such a purpose. With the collaboration of a half dozen colleagues, the Centro de Ecodesarrollo, sponsor of the project, published the results in three volumes. The Instituto de Investigaciones Sociales de la Universidad Nacional Autónoma de México laid yet another stone in this same road when it published some research carried out by a group of students from the Universidad Metropolitana and the Universidad de Yucatán on the corn-producing region of eastern Yucatan. I added one more pebble in 1982 when I prepared the inaugural exhibition, "Corn: The Basis of Mexican Popular Culture," for the opening of the Museo Nacional de Culturas Populares, brainchild of Guillermo Bonfil. That exhibit was the basis of a catalogue published by the museum. This may sound like rationalization or self-promotion. It may be rationalization in order to ensure that the original topic does not fall by the wayside, but it is not self-promotion. On the contrary, it is an acknowledgment of fellow researchers and colleagues who have shared my folly. I can state absolutely that the absence of Mexico in the present book owes to the fact that I am going to write another book specifically on that subject. One of the rules of the accumulation of information now seems clear to me: no one should learn more about a topic than they realistically can use to write a single book. I broke that rule and I will have to suffer the consequences.

Without realizing it, my interest in corn swept me up with uncontrollable passion. I gathered a significant amount of material on the history of corn worldwide. My access to the extraordinary libraries at Cambridge, the University of California at Berkeley, and the University of Chicago, libraries that take patrons seriously as researchers and not as petty criminals, was a surfeit, a banquet for a starving man. It also was a disappointment. Very few authors shared my unbridled passion. The search for materials began to slow down and I entered the phase of diminishing returns. Nevertheless, the material raised the possibility of a project I had not anticipated: a series of case studies on corn's role in the formation of the world system, nothing less than a world history of corn. The possibility both irresistibly tempted and terrified me. Temptation carried the day and my work led me to write a book in which Mexico scarcely figured.

Two preexisting conditions came to bear on this project. Like all Mexican social scientists, I am a Mexicanist by virtue of training, vocation, and manifest destiny. I know and study Mexico. The problems of that country define the horizons of my intellectual concerns. That is all well and good—it is one of the strengths of the Mexican social sciences— but it has its drawbacks as well.

At its most extreme, Mexicanism results in a lack of interest in what is happening elsewhere in the world, in spite of being conscious of the fact that the nature of our connection to the rest of the world defines and conditions our own circumstances to an important degree. More often, we simply take it for granted that knowledge about the world at large is a specialized field of knowledge that only the wealthy can aspire to, a luxury that we cannot afford. We end up resorting to Western scholars to enlighten us about those things that take place beyond our own borders. On the one hand, we become dependent on the acquisition and analysis of an essential portion of information in order to better appreciate our own circumstances. On the other, knowledge about the larger world remains characterized by the intellectual perspectives of developed nations. Those perspectives vary and the legacy of colonialism does not taint all of them, but colonialism does provide a context, does bring a certain reality to bear on all these perspectives. We do not enjoy global analysis that is at once a Mexican perspective and an international worldview. This abeyance leads to a distortion of knowledge about the world, and such intellectual unilateralism leads to scientific dependence.

It was not possible for me to renounce my Mexicanism. My research on corn held out the possibility of writing as a Mexicanist for a worldwide audience, of using my training and tools, my viewpoint, in order to take a look at a global issue. I imagine that I tended to favor certain processes and actors by virtue of my point of view, to see them in my own way. Nevertheless, to the extent that I was able I saw to it that my research did not lead to editorializing or censorship. The well-deserved criticism from some authors I consulted was a tempting prospect, but I chose not to risk falling into their analytical orbit and strengthening their prejudices, their imperial perspective. Reverse imperial history still would be imperial history. So, instead of disregarding what I saw, I tried to affirm it, to write a world history of corn with a Mexican twist. It sounds arrogant. Perhaps it is.

I owe the audacity in taking on a global perspective in great part to Angel Palerm, for whom nothing was sacred. He insisted on dealing with important problems, even though it meant being on equally intimate terms with Marx and Saint Ignatius of Loyola, or studying China, Alaska, or even a place as far removed as the Seychelles Islands. For Palerm, also first and foremost a Mexicanist, the world was a natural and logical setting. A global perspective held out interesting and serious issues and there was work to do. This book owes much to Angel Palerm. I would have liked for him to have read it.

The other precondition was that of anthropologist. In the nineteenth century, anthropology blithely speculated about universal evolution. In the twentieth century, the discipline became oriented intellectually and method-ologically toward knowledge about the small reality, a delimited area where everything was situated within walking distance and where everyone knew each other by name. I had experienced the intense satisfaction of the type of knowledge that such a practice tended to yield: profound, concrete, precise. Theoretical reflection was the product of contemplating concrete objects and known, real, specific persons. This type of knowledge also was the source of great frustration. Such a small reality gave us an insufficient perspective to explain many of the things that we saw there, things that had origins in wider, far-removed ambits. As reduced as the universe of study in this small reality might have been, we still perceived the presence of more worldly influences: the global market, modern science and cutting edge technology, national government, the roots of intrinsic dependency. By various means, we anthro-pologists tried to overcome the limitations of the small reality without losing the very integrity of our nature as a discipline, the very style that made our work unique. That is exactly what happened as we expanded our universe of study to include the surrounding region. We went on to study the nation state and once again, unfortunately, the wider world system appeared. We travelers and pedestrians, those of us who studied walking social relations that had first and last names, those of us who threw ourselves into our field work, to both the dismay and delight of our informants, whether we liked it or not, had the world system constantly before us as an object of study.

The selection of a familiar and concrete object such as the corn plant, something so close at hand, as a means to penetrate the world system owes much to Sidney W. Mintz and Eric R. Wolf, master anthropologists by trade. Their books have dealt with the world system without distancing themselves from the tradition or the type of concrete knowledge that is the hallmark of the anthropologist. They have opened a new niche for a book such as this. Once said, I would like to make clear that they bear no responsibility for any outrageous statements I may have made, nor can they be associated with any of my shortcomings. I say this with a certain advantage: I read them, but they are not familiar with my work here. I simply thank them for what they have taught me.

I owe a debt of gratitude to others for many and diverse reasons. I plun-dered the library of the agricultural historian Teresa Rojas Rabiela for infor-mation. In exhaustive conversations with her, I voraciously consumed every bit of knowledge she possessed on the topic. The Instituto de Investigaciones Sociales de la Universidad Nacional Autónoma de México, where I work, sponsored the project in hard times. While it was not able to be overly generous in financial resources, it did give unreservedly of those things that it did have at its disposal. How generous it was in granting me freedom, in bestowing respect and confidence, all of which were much more difficult to come by than funding. I owe thanks to many more who shall go unnamed, although not for a lack of gratitude.

It remains for me to explain the title of this book, a title that includes a word that many may consider excessively strong: "bastard." While we may chafe at the term, its use is entirely appropriate here. The absence of or disagreements over corn's progenitor transform such a strong word into nothing more than a descriptive adjective. I also use "bastard" in the sense of a person who has moved outside of his former social orbit, of one who remains outside the system of accepted norms. Corn entered the world system in just this way. Enlightened elites used corn in this sense: as a contemptible object subject to discrimination. Corn carried the stigma of being alien, strange, poor. The wealthy judged corn and declared it to be guilty. The poor, on the contrary, opened their doors to it, embraced it, and adopted it. Corn shared the fate of the poor, of those of mixed race, of the unchaste. And corn thrived virtually everywhere. Corn was an adventurer, a settler of new lands, one of those that helped fashion the modern world from the distant sidelines. Corn was nearly absent from the colonial metropolises attributed with the construction of the modern world. Corn was on the frontiers of the modern world, from where the modern world effectively sprang by virtue of hard work, imagination, and innovative irreverence. Like many stories about bastards, this one has a happy ending. Corn's true identity as a ruler of the Western world is now a fact. Such a happy ending is not the final episode. Rather, it is only the beginning. Now the bastard reigns. We hope that corn does so with justice, with grandeur, and a desire to serve. Corn came into its own hand in hand with the poor of this world, a lesson it would do well not to forget. The bastard king can be one of the champions in the struggle for a world without hunger. It is something that corn owes to its past, to its history. It is something that we can accomplish.

Corn & Capitalism

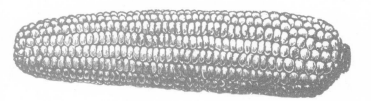

AMERICAN PLANTS, WORLD TREASURES

Vast riches, born of violence, flowed freely between the New World and the Old since the time of discovery. Those treasures took many and diverse forms. Precious metals and other highly valuable New World commodities such as dyes generated large amounts of ready cash. Plentiful American land and labor allowed for the production of those goods that were craved and coveted in Europe: sugar, coffee, and a whole array of other plant and animal products. The New World provided an open frontier for the Old World's surplus population of undesirables: the poor, the persecuted, the fanatics, the heretics, the bureaucrats, the adventurers of every ilk. America was a place to realize dreams that were otherwise unimaginable in Europe. Successive waves of the hopeful, renewed and reinvigorated, departed to seek their fortunes in the New World. America's plentiful and unique natural resources transformed life, production, and their nexus. The New World generated immense new markets that willingly or forcibly trafficked in everything from European manufactured goods to African slaves. Those markets generated vast resources that then were used to expand and dominate world trade. In the end that wealth, transformed into capital—into a relation of production and of ownership—was a central element, if not the primary element, responsible for uniting many local economies into one immense world market and in the formation and development of capitalism as a hegemonic global force.

In the five hundred years since contact between Europe and America, plants have stood out among the many treasures discovered in the New World. The wealth generated by plants probably has increased at a greater rate and in a more sustained fashion than any other American resource. In any given year—1980, for example—the annual value of American crops, on the order of $200 billion, probably is higher than the total value of all the precious metals exported from the Iberian colonies over the course of the entire colonial period. The seven most important food crops today—wheat, rice, corn or maize, potatoes, barley, sweet potatoes, and cassava—supply at least half of all nutrients consumed worldwide. Four of those plants are from America—corn, potatoes, sweet potatoes, and cassava—and make up half of the total volume of

1

the top seven crops. More than a third of the modern world's food, either fresh or processed, comes from American plants (Harlan, 1976). American plants are a potential source of great wealth, but also of great poverty, misery, and exploitation.

Native American plants came to have a wide range of uses in the Old World. Europeans regularly used the Xalapa or Mechoacan root (*Ipomoea purga* [Wenderoth] Hayne) in the sixteenth and seventeenth centuries as a powerful laxative. The Spanish physician Nicolás de Monardes was responsible for popularizing this Mexican plant. De Monardes wrote a work on American medicinal plants that was published in 1574 and was well received. During this same period, brightly colored fabrics using dyes extracted from cochineal or Brazilwood were a sign of conspicuous consumption. Cochineal was extracted from an insect native to Mexico, while Brazilwood or peachwood (*Haematoxylum brasiletto* Karsten) was a tree native to the American humid tropics. Cinchona bark (*Cinchona* species), from a tree native to the Amazonian Andes, was used in a preparation that effectively prevented or treated malaria. This treatment made possible the settlement of vast areas of humid, swampy land the world over (Hobhouse, 1986: chap. 1). Cinchona bark was used in making gin and tonic, the colonial era's signature beverage. This drink had the virtue of being both intoxicating and medicinal, meeting two pressing needs among colonial enclaves. A wide range of native American plants were used for medicinal purposes long before drugs were artificially produced in the laboratory.

Today, native New World plants continue to be enormously important. Cortisone was produced for many years from *cabeza de negro* (*Dioscorea mexicana* Guillemin), a plant native to coastal Mexico. Medical advances associated with cortisone may be comparable in significance to those resulting from the discovery of penicillin. Barbasco (*Dioscorea composita* Hemsley) was used to make the birth control pill. Mexican peasants gathered the plant along the coast, where the plant was indigenous. Without the widespread availability of the pill, it is impossible to explain the sexual revolution of the 1960s that forever changed the way of life and culture in industrialized countries and in certain sectors of underdeveloped countries. The impact of such hormonally based contraceptives also was linked to demographic patterns, some of the most complex and far-reaching trends in modern society. Rubber became indispensable at the outset of the Industrial Revolution. Originally, the resin used to produce rubber was obtained exclusively from American plants, especially from a tree native to Brazil (*Hevea brasiliensis*). Synthetic forms of rubber were developed during World War II. Some decades later,

American Plants

resins used to produce rubber also were obtained from guayule (*Parthenium argentatum* A. Gray), a wild shrub native to the arid regions of northern Mexico and the southwestern United States. This use of guayule is undergoing something of a revival today and the production of natural rubber is on the rise as well (Consejo Nacional de Ciencia y Tecnología, 1978). Precious ornamental woods, other resins, and many pharmaceutical products are part of the extensive repertoire of native American plants.

Plants native to the Americas have a great deal of potential. This is especially true as the world finally comes to grips, economically and intellectually, with the indisputable and unpleasant truth that petroleum is a nonrenewable resource. Plants, on the other hand, are capable of both reproducing and multiplying. Plants native to arid regions of Mexico have become fashionable of late. These include guayule, jojoba (*Simmondsia chinensis* Link), catkin (*Euphorbia cerifera*), gobernadora (*Larrea divaricata* Cavanilles), and peyote (*Lophophora williamsii* Lemaire). A wax that substitutes for whale blubber in the cosmetic industry is made from jojoba. Another type of wax is extracted from catkin. *Gobernadora* is used to produce a substance that retards the oxidation of fats and oils. In short, American plants show tremendous promise. Even the history of plants native to the Americas is subject to irony, however. One only need conjure up an image of the typical U.S. tourist, the quintessential Ugly American, methodically chewing gum extracted from the sapodilla tree (*Achras zapota* Linnaeus), a tree native to the Mayan jungle.

Some of the American plants that have changed our lives and the very course of history were wild, natural products of the gradual evolution of flora on the American continent. People did not play an active part in the appearance of these wild plants. They did play a role, however, in the both the preservation and, in some cases, the extinction of those plants. People also obviously used their knowledge of plants' traits and properties to establish and define the uses of indigenous plants. How native wild plants are used is a product of culture. Daily life and medicine, both past and present, would be very different without indigenous American plants and the knowledge that New World native peoples accumulated in order to exploit them.

The history of the myriad and at times surprising present-day applications of indigenous American flora is a work in progress. Their uses are not the result of happenstance. Neither are they attributable simply to the genius or good fortune of gifted scientists or inventors. Current medicinal or industrial uses are, generally speaking, extensions and adaptations of age-old collective practices and knowledge. Thomas Adams, for example, widely acknowledged as "the king of chewing gum," made a fortune simply by adapting and commercially marketing an indigenous American habit. Likewise, the developers of patent medicines capitalized on discoveries made by those they scornfully referred to as "witch doctors." While indigenous knowledge has been dismissed by Western standards as merely empirical, the cultures that created such knowledge have in fact been systematic in preserving and utilizing their findings. Indeed, written accounts several centuries old have occasionally turned up. None of this detracts from the imagination or commercial genius of Adams or other modern entrepreneurs. It simply recognizes that private parties have at times appropriated collective, historic, and public knowledge.

Indigenous cultures had a profound, systematized knowledge of nature, which is the basis for the past and present uses of American plants. What is perceived as weak, mechanical development in such indigenous societies is often used as a justification for classifying those cultures as savage or barbaric. The depth and complexity of systems of knowledge related to the uses of indigenous plants, however, would seem to amply compensate for any such perceived weaknesses. This outstanding strength in the natural sciences, especially in the realm of genetic engineering, led to the creation and exploitation of a vast repertoire of renewable resources.

The most singular example of the creation of such plant wealth, of biological capital, was the extensive inventory of New World plants exploited and cultivated before contact. Domesticating plants, subjecting plants to human labor through cultivation, was a long process everywhere, extending over thousands of years, during which plants were made to serve the needs of people. Groups of people, for their part, adapted to agricultural life and met the challenge of producing surpluses over and above the minimal cultural and nutritional necessities of the cultivators alone. Thus, permanent settlements and sedentary life arose; cities appeared and with them specialized groups that produced, administered, and governed without concerning themselves with producing their own food. Complex societies developed and classes, ranks, and specialized professions emerged within those societies. The complex process of domestication implied the collection and accumulation of knowledge about plants and plant traits, about factors determining their growth and reproduction, and about the capacity of determined social organizations to direct and organize the development of the plant world. Land, water, temperature, winds, seasonal considerations, heavenly bodies, clouds, mountains, agricultural practices, organization of labor, and the preparation and long-term storage of food became objects of systematic observation, analysis, experimentation, correlation, and explanation. Plants were selected and transformed gradually but dramatically, methodically separated from their wild

American Plants

ancestors under the careful direction and observation of people in order to better serve human needs. The domestication of plants implied, in short, the accumulation of knowledge.

The development of agriculture was neither a universal process nor an obligatory phase in the evolution of all human groups. On the contrary, it was something that happened in very few places and as an exception to the rule. According to Nikolai I. Vavilov (1951), in his modern classic on the origin of cultivated plants, only eight primary centers for the domestication of plants have been documented. From these, agriculture extended outward, whether by example, imitation, or conquest. There were two primary centers for the domestication of plants in the New World, these only recently acknowledged and still subject to dispute, and secondary centers arose around these. Together, these agricultural centers contributed more than a hundred new plant crops to the indigenous American repertoire, a number equal to half the entire agricultural heritage of the Old World (Harlan, 1975: 69–78). Those who domesticated American plants were dealing with an entirely different form of vegetation and extreme variation in environmental conditions as compared to the Old World. These early New World cultivators not only created new crops but also new techniques for production, intercropping, storage, and consumption. In America, new agricultural cultures arose that allowed the development of highly varied and complex civilizations. Above and beyond their mastery of agriculture, these civilizations left behind ample evidence of their knowledge, their experimental boldness, their nonintrusive mastery of nature, and so many other things that made up a valuable heritage, a legacy only barely acknowledged by subsequent generations.

The diets of the large pre-Hispanic population were almost entirely vegetarian. Foods of animal origin were only modestly represented. They were not intentionally vegetarian as we think of it today, but rather they were extremely frugal in the consumption of meats. There were very few domesticated animals in the New World. Nature, so lavishly bestowed with plants, was far less well endowed with edible animals: the humble and prized turkeys, ducks, dogs, rodents, Andean ungulates—llamas and alpacas—bees, and cochineal, that is, if one can speak of the domestication of insects. Of these, only the first five were edible. There was some hunting of wild animals and gathering, including many varieties of insects. These complemented or supplemented what domesticated animals there were in order to supply the animal protein necessary for human nutrition, which in any case was very low. All their other food came from plants. There was no evidence of any severe nutritional deficiencies or hardships.

Among the one hundred plants New World inhabitants cultivated, there were plants that were quintessentially sweet, fatty, and spicy hot. Many of the cultivated American plants gained notoriety and importance after contact, while others still today are used only locally. Some basic foodstuffs stand out among the plants that came to have enormous importance in the making of the modern world: those foods that were an integral part of a daily diet, such as tortillas, bread, and rice, an essential part of all meals rather than a side dish. Sixteenth-century Spaniards called them, more precisely, maintenance foods. This type of food provides the greatest share of daily calories, the energy to sustain human activity, and other essential nutrients. It is with maintenance foods that we most often associate the idea of hunger or its counterpart, a sensation of being full and satisfied. These basic foodstuffs are the dietary foundation of any meal. Maintenance foods are the stuff of a balanced and adequate diet, of a delicious meal, of a gratifying and pleasant gastronomical experience. Maintenance foods distinguish dining apart from the simple act of nourishing ourselves. Thanks to maintenance foods, a meal becomes something special.

Corn, or more properly maize (*Zea mays* Linn.), and potatoes (*Solanum tuberosum* Linn.) stand out among the basic American subsistence foodstuffs as tremendously important. It is impossible to understand the agricultural revolution that spurred both demographic growth and accelerated urbanization, making the Industrial Revolution possible, without considering the impact of corn and potatoes. The tragic Irish potato famine dramatizes the history and importance of the potato. Historians have documented that tragedy in which between eight hundred thousand and a million people died between 1845 and 1851—out of a total population of seven and a half million—when potato smut destroyed their basic food, the potato (Salaman, 1949; Grigg, 1980). Corn has received less attention, although attempts certainly were made to introduce it into Ireland in order to avoid a repetition of such a tragedy.

Corn, cassava (*Manihot esculenta* Crantz), and the sweet potato (*Ipomoea batata* Linn.), American maintenance foods, are crucial for understanding the settlement and subsequent demographic growth of tropical regions the world over, a process that has received very little attention from researchers. It is impossible to explain the accelerated growth of the world's population since the eighteenth century without taking into account American maintenance crops and their growing productivity. Population growth in burgeoning regions has not abated even today and for many nations constitutes a burden for those least able to tolerate it, especially among people of color.

Some basic foods of the American past initially lagged behind in importance but remain a potential source of promise for the future. Among these are quinoa (*Chenopodium quinoa* Willdenow), native to the Andes and noteworthy for its high protein content. The same is true for sesame (*Amaranthus cruentus* Linn.), also known as *alegría* or "joy" for the comfits made from its grain and unrefined sugar cane. This plant, widely eaten in pre-Hispanic Mexico, fell into disfavor during the colonial era when the conquistadors subjected it to systematic persecution by virtue of its association with indigenous religious cults. The labor intensive nature of its harvest, incompatible with the severe population losses in New Spain, also contributed to its decline. Recently, sesame has come to the attention of agronomists and nutritionists alike for its elevated content of high quality protein. Some consider it a perfect source of protein, and today research and experiments are being conducted to increase its productivity and to farm it on a commercial scale (Cole, 1979). Researchers' ongoing efforts indirectly highlight how peasants in Mexico and elsewhere, such as India, have preserved this genetic patrimony over several centuries as part of their agricultural culture and nutritional insight. Sesame, a resource of some renown in the past, can provide hope for the future as an unexpected but welcome source of alternatives.

Other American plants have become important the world over since contact. Their impact has been less apparent than the overwhelming quantitative impact of maintenance foods, but not insignificant or inconstant. Few could imagine modern European cuisine without the tomato (*Lycopersicon esculentum* Miller), originally from Mesoamerica. Other plants might provide the same nutritional value as the tomato, but no other plant can match its flavor and texture. By now the tomato is such an integral part of European cuisine, even thought of as a traditional ingredient, that the revelation of the tomato's exotic origin would surprise and perhaps even offend many Europeans. The chili pepper (*Capsicum annuum* Linn.) also has made a place for itself outside the Americas, less so in European cuisine than in Hindu and Chinese dishes, where the tomato also figures (Long-Solís, 1986). It is not so easy to substitute for the nutritional value of the chili pepper. This is not so much because of the chili's nutritional value as determined by Western standards, but rather for precisely what Western tradition rejects: the hot, spicy quality of peppers, which makes for the more efficient absorption of other nutrients, although this aspect has not received much attention. Much could be said about frijoles (*Phaseolus* spp.), the ideal complement to corn by virtue of their elevated protein content. Frijoles, just like chili peppers and squash (*Cucurbita* spp.), were consumed all over pre-Columbian America. They slowly made their way around the globe until they became an indispensable ingredient in the most distant and remote places.

Nopals (*Opuntia* spp.) and magueys (*Agave* spp.), originally from Mexico, are now part of rural landscapes half a world away. Neither pulque nor tequila can be extracted from the maguey that grows on the Mediterranean rim, a reminder that some traits are subject to loss as plants migrate from one place to another. Cochineal, the insect cultivated for its pigment in ancient Mexico, cannot be raised in the nopals abroad, except in the Canary Islands. Neither magueys nor nopals are without their redeeming qualities, however. These plants serve to effectively stem soil erosion, something the ancient Mexicans also discovered, hence justifying their introduction and use. The list of little-known peculiarities and trivia can go on and on and even raise important issues in appreciating how different cultures interact and how that interaction brings about change. We call such change, at times refracted through a moralistic and prejudicial prism, progress.

Ancient American crops provide many modern luxuries. Tobacco (*Nicotiana rustica* Linn.), now in disrepute for its health risks, was at one time an extravagance that few could afford, only later to become a proletarian pleasure in industrial societies. Native peoples have cultivated and smoked tobacco since pre-Hispanic times, as attested to by the astonishment of the expeditionary accompanying Columbus upon seeing the natives "eat smoke." The ancient Mexicans so valued the cacao bean (*Theobroma cacao* Linn.) that they used it as a medium of exchange. The inhabitants of ancient Mexico drank chocolate made from cacao beans long before it was considered more Swiss than Swiss watches, a special treat for well-behaved children and the bane of the obese. Vanilla (*Vanilla planifolia* Andrews) is an essential ingredient for bakeries in wealthy nations and for the wealthy the world over. Vanilla provides the true sum and substance of ice cream. Native peoples still cultivate the bean on the coast of Veracruz, just as they did thousands of years ago, despite the development of an artificial substitute that makes imitation vanilla readily available. Cochineal has been displaced by anilines, synthetic dyes of lesser quality. Today, cochineal dyes are reappearing in indigenous weavings in Oaxaca.

Other luxuries were never as cosmopolitan, but no less important. Copal (*Bursera jorullensis* Engler) and other aromatic resins served as substitutes for incense in America. The fragrance of copal permeated church sanctuaries, just as in pre-Hispanic temples, and wafted into the religious recollections of their worshipers. Annatto (*Bixa orellana* Linn.) can be used for red dye and imparts a distinctive flavor to food, and also served as a substitute for imported saffron. The list of clandestine luxuries, secrets of the poor peasants of America that make their festive cuisine one of the world's most varied, could go on and on.

Until relatively recently and with few exceptions, New World inhabitants, especially those living in the tropics, were the only ones to enjoy American fruits. The situation has been slow to change and now the avocado (*Persea americana* Mill.) and the guava (*Psidium guajava* Linn.), together with the pineapple (*Ananas comosus* Linn.) and the papaya (*Carica papaya* Linn.), all of American origin, are part of an array of international delicacies. Sapodillas and mammees (*Calocarpum mammosum* Linn.), soursops and many other fruits, too perishable to transport and adapt, lend themselves more to regional consumption. They now compete with Old World fruits, also widely cultivated in America.

Oil is a by-product of two American plants that have been widely disseminated the world over: the sunflower (*Helianthus annuus* Linn.) and the peanut (*Arachis hypogaea* Linn.). Both figure on a list of the thirty leading crops in the world today, as do frijoles, tomatoes, tobacco, cocoa, cotton, and of course the four American maintenance foods. Coca should figure conspicuously on this list, but the available figures are erratic, ambiguous, and unreliable. They are compiled or concealed in police departments rather than in agriculture departments.

Coca cultivation (*Erythroxylum coca* Lamarck), originating in the Andean piedmont, dates back a thousand years. Inhabitants of the Andean region chew the leaves of the shrub, a practice integral to Andean culture. It is not easy to characterize the use of coca using Western categories. Medicinal, religious, ceremonial, prophetic—all these concepts seem inadequate, describing only limited aspects of the multiple and complex purposes that coca serves in Andean life. Coca is all this and much more. Coca is a legitimate and essential resource for that culture, a culture understood as an ensemble of means to relate to social and natural media in the extreme conditions of the Andes, where human settlement took place at higher altitudes than in any other part of the world.

So important was coca that the conquistadors transformed it into a major business venture. The monopoly to sell coca was almost as profitable for Spanish mining interests as the extraction of ore itself. In the nineteenth century, scientific research found a way to extract an alkaloid concentrate of the coca in coca leaves. The Western world accepted and even extolled coca's medicinal and commercial use. Coca was one of the ingredients, omitted today, in the original formula for Coca-Cola and was probably one of the factors in that beverage's immense popularity. Other widely marketed tonics used cocaine liberally before public health officials prohibited or restricted its use.

The illegal traffic in cocaine is one of the most important phenomena in the last quarter of the twentieth century. There are millions of casual or habitual users in the First World. That world is characterized by alienation, weariness, and escapism, all of which work to create a market for illegal drugs of enormous proportions. The value of cocaine perhaps exceeds that of any other Latin American export product. Supplying the cocaine market produces contradictory and complex processes in areas where coca is cultivated. The settlement of vast regions, the survival of the peasant and in some instances of entire countries, cannot be accounted for except by the cultivation of coca. Repression, armed violence, corruption, extreme inequality, chronic crime—these are the trade-offs. The societies affected by the cultivation and consumption of coca have not been able to confront the complex phenomenon beyond moralistic discourse and repression that, typically, deals harshly with the downtrodden while protecting the privileged. Those directing the drug traffic are rewarded with impunity and profit while dangerous defoliants lay waste to bountiful agricultural lands and peasant economies. Coca is an unfinished and painful chapter in the story of American plants.

Cultivated American textile plants are equally important. Henequen (*Agave fourcroides* Lemaire) was originally cultivated by the Mayas living in the Yucatan peninsula. Its sisal variety was introduced in Africa once its commercial value and importance were fully appreciated. Henequen was enormously important to the role sailing vessels played in advancing navigation. Later, henequen had a complex role in the history of the Industrial Revolution. The rope and twine made from henequen spurred the growth of agricultural mechanization. Henequen figured prominently in the Caste War of Yucatan, which resulted in the foundation and functioning of an independent Mayan state in Quintana Roo for half a century (Benítez, 1986; Reed, 1964).

Cotton (*Gossypium* spp.) was one of the few plants that undoubtedly was cultivated in both America and the Old World for many years before contact. The fact that the species cultivated in America—*hirsutum* and *barbadense*—were different from those harvested in the Old World suggests that it was a matter of independent and parallel domestication. The American varieties had some important advantages over their Afro-Asian cousins, as inferred from the fact that virtually all modern commercial cotton varieties are derived from the American varieties. Even renowned Egyptian cotton, highly valued for its quality, is a variety of the Barbadian species of American origin. It must have been a fascinating story, that of the migration and replacement of one species of the most important textile plant in the world for another, a plant that also became an important source of oil. Once again, the Industrial Revolution figured in that tale, for cotton fabric was the first large-scale exportable commodity and the textile industry was the motor of industrialization.

There are many other cases exemplifying the repercussions of New World cultivated plants, the development of the world economy, and the world marketplace. The examples only hinted at here are part of that larger tale but, at the same time, each one is an ensemble of many other stories spanning time and space. There are far too many episodes to explore and recount here. It is preferable to select only one to examine more closely. Hopefully, this will allow us to better understand the process, the players, and the ensemble of fateful forces and accidents. It will give better insight into that which was, that which might have been. This is the story of corn, the protagonist of this book. At the heart of this drama, this comedy, in which corn figures so prominently, lies the history of capitalism.

: 2 :

BOTANICAL ECONOMY OF A MARVELOUS PLANT

The corn plant is a gigantic annual grass belonging to the family *Gramineae*. It forms part of the subtribe *Maydeae* that has five genera, three American and two Oriental, and is the only species of the genus *Zea*. In scientific nomenclature it is known as *Zea mays*, a name given to it by Carolus Linnaeus—founder of the system of binomial nomenclature for living things—in the first half of the eighteenth century. The name *Zea mays* comes from two different languages. *Zea* comes from ancient Greek and it apparently is the generic designation for grains and cereals in general. Other sources have suggested another meaning for *Zea*: that which gives and sustains life. *Mays* probably comes from Taino, the language spoken by aboriginal groups in the Antilles, where Europeans first stumbled upon the plant. The term *mays* has another meaning as well: life giver. If these translations are accurate, corn's scientific name is perfectly redundant.

In 1591 Juan de Cárdenas, a doctor from Seville residing in Mexico, published a medical book entitled *Problemas y secretos maravillosos de las Indias*. In it he wrote in an awestruck and alluring tone: "Corn is one of the seeds that might better be known for the esteem in which we should hold it in the world, and this for many reasons and causes" (Cárdenas, 1988: 171). Four hundred years later, this allure has been fulfilled: corn is known, renowned, and used the world over. Corn harvests exceeded 440 million tons a year by the late 1980s. Worldwide, corn was the third largest crop by volume, just behind wheat and rice, and sometimes drawing even with the latter. Its rate of growth was higher than that of other cereals, which probably will make corn the most important crop in the new millennium.

Let Juan de Cárdenas, speaking to us from the sixteenth century, be our guide in exploring corn's universal appeal. Among corn's virtues, Cárdenas notes: "First, it is widespread, that is to say that it is a seed that in cold, hot, dry or wet climates, in mountainous regions or pastures, as a winter crop and as a summer crop, irrigated or dry farmed, corn is cultivated, harvested, and thrives" (Cárdenas, 1988: 171). Corn displays a tremendous capacity to adapt to diverse and even extreme ecological conditions. Corn is enormously

12

flexible. At the time of contact between the New World and the Old, corn was cultivated from forty-five degrees latitude north, as far north as Montreal, to forty degrees latitude south, over six hundred miles to the south of Santiago (Weatherwax, 1954: chap. 7).

This long imaginary line, running more than six thousand miles, passes through a variety of contrasting conditions: arid deserts, tropical rain forests, lofty mountain chains, enormous temperate plains with frigid winters, the highest altiplanos in the entire world. All in all, corn cultivation ranges from sea level to heights of almost ten thousand feet.

The corn plant requires, on average, a period of around 120 frost-free days with good sunlight for a crop to mature. Corn is a tropical plant and its poor resistance to cold weather is one of its weak points. On the other hand, it takes better advantage of sunlight and grows more rapidly because of the size and disposition of its foliage. The wide, exposed surface area of the corn leaves efficiently captures the sun's rays. For this reason, corn sown far from the equator is exclusively a summer crop. Lack of rain affects corn most at two critical periods: germination and early development and pollination. Tasseling takes place in most commercial varieties about a hundred days after sowing, although there are varieties that mature even earlier. Corn does not require excessive watering, although wheat and temperate zone cereals require somewhat less. Corn does require watering that is evenly distributed over its growth cycle, consistent with the phases of the corn plant's growth. Dry-farmed corn has been documented in areas ranging from less than ten inches to nearly two hundred inches of average annual precipitation. Corn's vast geographical breadth includes regions where corn is cultivated under irrigation in order to guarantee sufficient moisture during the critical periods of its growth cycle. Where the rains are more or less regular, growers dry-farm corn by relying only on seasonal rains.

Frost and drought are the phenomena that determine corn's natural geographic limits. But it is a way of life, and culture, that brings such natural potential into the present and at times forces it beyond its former limits. Before contact, corn cultivation and the very practice of agriculture were unknown in Argentina's humid pampas, today one of corn's most highly productive regions. Neither was corn cultivated on the great plains of North America, a leading corn-producing region today. The peoples that lived in those areas lived primarily from hunting. In contrast, the mountainous slopes, narrow valleys, and glens of the great subtropical mountain ranges, which today are only statistically marginal producers, were formerly the leading regions for corn cultivation. The peoples that settled and developed in those areas not only created complex systems to transport water, but also to move the soil, in order to mold and even to create land so that it served to sustain the cultivated plants on which they built their civilization. There is a complex interaction between cultures and natural frontiers in the grand design of agricultural geography.

The great variety of conditions under which corn could be cultivated were expanded even further after corn migrated abroad. Today, more than fifty countries on six continents sow corn in areas totaling nearly a quarter of a million acres. Among the ten most important corn-producing countries today, only four are on the American continent, corn's native soil. Today, corn's patrimony is universal.

Corn's enormous adaptability is partially attributable to that plant's physiological and physical characteristics. It is even more attributable to human labor and knowledge. Even though corn is only one species, it has a great number of races and varieties that differ enormously. This is reflected in plant size, which can range from three to thirteen feet in height. The period between germination and pollination can vary from 45 days to more than 150 days. It is possible that sometime in the past there were varieties that matured even sooner. The number of leaves on a corn plant can vary from eight to forty-eight and the number of ears, although almost always one per plant, can be double or triple that. An ear of corn can measure anywhere from four to twenty-four inches long, and the size of the kernels can vary tremendously. Then there is the matter of the colors of the kernels, which are of four basic hues: white, yellow, red, and purple or black, seen in all shades and combinations. There are six corn varieties for commercial purposes: flour corn, waxy maize, sweet corn, popcorn, flint, and dent. Each has a different starch content and texture and satisfies a different type of demand. There are even more varieties that do not enjoy a wide commercial market. These and other aspects of corn's variability serve to make corn adaptable to diverse environmental conditions and myriad human needs.

The majority of corn races and varieties can be categorized as either accidents of nature, such as natural crossing or free pollination—associated with corn's pollen-producing capacity that greatly surpasses that of other cereals—or genetic mutations—to which corn is especially inclined. The selection and preservation of such serendipities and their specialization according to the potentialities and limitations of the environment is the result of human intervention, of agricultural knowledge and its accumulation. Scientists have devoted monumental efforts to compiling a complete inventory of existing corn races and varieties. This work has at times been complicated by a lack of consistency in the criteria used to classify and name corn varieties. Even so, specialists have identified more than 250 corn races. In Mexico alone, specialists have identified at least twenty-five antique corn races with hundreds of varieties (Wellhausen et al., 1952). That genetic treasure, that remarkable plant heritage, not only is important in order to explain the enormous flexibility that so impressed Doctor Cárdenas, but constitutes the basis for developing even better corn varieties in the future. That historic accumulation, starting with what most certainly is an exceptional plant, growing almost everywhere there are people, justifies labeling corn as a miracle plant.

Let us return to Doctor Cárdenas and other virtues of corn: "Second, for its abundance, which is to say anywhere from one hundred to 200 measures of grain per measure of seed sown, and this without too much work, but easily and leisurely, not waiting almost from one year to the next, like one waits for wheat in Spain, which is sown in October and picked in June or July, something corn does not do, for within three months and at most four, and in some regions within fifty days, it is picked and stored" (Cárdenas, 1988: 171). Other chroniclers of the era note even higher yields for corn, as high as 800:1, but the typical yield most often mentioned by the chroniclers of the Indies is that of 150:1. Nowadays, we would speak of an elevated yield in combination with a short land use cycle.

At the time Doctor Cárdenas wrote, agricultural yield was measured by the ratio between the amount of seed sown to the harvested crop brought in. Many peasants use this same sort of reckoning even today. That system of measurement most likely assumed abundant land and scarce labor. The labor necessary to sow seed was the crucial variable in calculating a rise in productivity or a higher yield. Dutch agriculturists had the highest yields in Europe in the first half of the sixteenth century. They brought in, on average, 9.7 units of wheat for each unit of seed sown, 9.1 units of rye, 7.4 units of barley, and 5.4 units of oats. Elsewhere in Europe, yields of 6:1 or even 4:1 for wheat were considered customary and acceptable. Two hundred years later, between 1700 and 1749, the yields of traditional European cereals had fallen by 20 percent with respect to the sixteenth century (Maddalena, 1974; Slicher van Bath, 1963: Appendices). As far as growing corn is concerned, it is important to keep in mind that the number of individual corn plants that could be grown per unit area was much lower than the number of individual Old World cereals on the same sized plot. Even so, at ratios of 150:1 in typical years, and even 70:1 in bad years, figures for Mexican corn yields were certainly remarkable (Humboldt, 1972). It is also important to keep in mind that while Old World cereals could be sown once every two years in the Mediterranean and twice every three years in continental Europe, in many parts of America corn was sown two and sometimes even three times a year. These were two entirely different worlds, indeed.

Times changed and so did the method for calculating agricultural yields. In the nineteenth century it became standard practice to reckon yields in terms of a ratio of the harvested crop to the sown land area. This new way of measuring indirectly reflected the relative scarcity of land in times of an abundant labor supply, which was the result of explosive demographic growth beginning in the eighteenth century. Corn continued to lead other cereals, even using this new method to calculate agricultural yields. In 1982, a typical year, the international average yield was 1.5 tons per acre for corn, just over 1.3 tons for rice, 0.92 tons for barley, and 0.91 tons for wheat. The United States was the world's largest corn producer and the world's second largest wheat producer during the 1980s. The United States used the best available technical resources for raising both crops, but only as long as it remained economically feasible to do so. Even with this stipulation, corn yields overwhelmingly surpassed those for other crops. The average yield for corn in the United States in 1982 was just over 3.2 tons per acre as opposed to just over a ton for wheat. The average yield for corn in the United States easily surpassed all other average yields for the cultivation of cereals elsewhere, including rice in Japan, at around 2.5 tons per acre, or wheat in England, at 2.8 tons, or in Holland, at around 3.1 tons (Food and Agriculture Organization, 1983).

Corn's high productivity is, in the last analysis, a reflection of the high photosynthetic efficiency of the plant, of its capacity to transform solar energy into living matter. This is a direct result of the large surface area of the corn plant's foliage exposed to the sun and its ability to transform light as well as perform other complex chemical processes. The architecture of the corn plant partially explains its capacity for growth, which is outstanding when compared to other cultivated plants. Corn transforms light, heat, and other inorganic elements into biomasses with the least amount of waste. The plant actively responds to the transformation of solar energy, the most abundant of the inorganic elements, and becomes a resource readily available to people.

Corn's high yield is tied directly to the surprising fact, unique to the gramineaes, that many kernels are concentrated on an ear, the husks acting as a novel protective covering for the entire ear. Other cereals cover individual grains with bractea. An ear of corn is a true marvel of order and symmetry, whose architecture is both utilitarian and beautiful. An ear commonly has more than 300 fertile seeds neatly aligned in rows. Five hundred is an average number of kernels on those ears I have examined and it is not unusual for this number to be as high as a thousand. These perfectly ordered kernels are the product of one lone seed. Corn kernels are much larger than those of other cereals, yet another factor in corn's high yields. Some kernels are so large, such as those of a Peruvian variety native to the Cuzco region, that they measure anywhere from three-quarters of an inch to well over an inch in diameter and are eaten individually. The high number and the concentration of corn kernels is only possible because the ear occupies a low, central position on the plant stalk, which allows it to capture a greater proportion of nutrients. This is a characteristic unique to corn as compared to other cereals, whose seeds are placed in a high, lateral position, making them much less resistant to weight and causing them to receive fewer nutrients. The ear, truly unique in nature, is a product of the corn plant's enormous capacity to concentrate energy in the kernels, in the fertile seeds. More still, the ear reflects the accomplishment of the peoples who domesticated and developed that ability to convert energy into forms that respond to human needs.

The ear has other practical advantages for producers. The bractea or husk protect the corn from moisture once the kernels are mature. The husk also serves to shield the kernels from pests and from any accidental dispersal and loss of seed in the field or in transport. Picking ears of corn is relatively easy and implies only minimal waste or loss compared with other smaller grains growing on spikes or panicles, like wheat or sorghum. An ear of corn is readily separated from the stalk and the husk, while Old World cereals require threshing. The natural protection afforded the ear by its husk facilitates storage. From the point of view of corn management, the ear is a natural package, a secure and uniform type of container, gift wrapping if you will.

The corn plant has an extensive root system that permits it to draw moisture and nutrients from a wide area. This is yet another factor in corn's high yields. The extensive root system demands relatively wide spacing between plants, which is why there are fewer plants per plot as compared to other cereals. This apparent disadvantage is amply compensated for by the high productivity of each individual corn plant. The greater distance between corn plants also has a number of advantages for the cultivator. The distance between plants and the low density of plants per plot is precisely what permits corn to be planted and cared for individually, plant by plant. This would not be viable or reasonable with other cereals planted with greater density, sown by broadcasting. The sowing of individual corn plants means that tilling or plowing of the soil is not necessary, although it can be sown in this way. It is only necessary to drop the seed into individual holes in roughly plowed ground. Using no-till, corn can be sown on very steep hillsides with minimal risk of soil erosion or in very rocky soils where tilling is not possible. It is ironic that no-till, an age-old traditional method for sowing corn, has been hailed as one of the most important innovations of modern scientific agriculture. The possibility of sowing corn on lands that would not be suitable for the cultivation of cereals is, in the last analysis, another way of raising soil productivity.

The separation of individual corn plants makes sowing in regular rows with generous spacing between rows feasible. This has important advantages in systems in which the soil is plowed as well as in no-till systems. The distance between the plants and individual rows permits weed control once the corn has sprouted. This can be done by hand using a hoe or by the use of mechanical cultivators that uproot any competing vegetation. Such wide spacing is not characteristic of other cereals—with the exception of sorghum—which are so closely planted that they do not allow people, beasts, or machines to penetrate the field after sowing. The possibility of selectively eliminating competing vegetation by mechanical means without affecting the corn crop itself means higher yields due to a better exploitation of sunlight and nutrients. There also is less of a risk of interfering with the growth of the already sprouted plants. The history of agricultural mechanization is intimately tied to the opportunity corn offers for producers and farm implements to intervene in the development of the young plants.

The distance between corn plants favors the exploitation of idle land for agriculture. Corn can grow simultaneously and complementarily with other useful cultivated plants. When this takes place, the total yield of the land is much higher than that of the corn harvest alone. In ancient Mexico, corn frequently was intercropped with squash (*Cucurbita* spp.) and frijoles (*Phaseolus* spp.). Squash is a creeper that does not compete with corn for light. Its foliage and the shadow it casts restrict the growth of weeds and lowers the evaporation of ground moisture. The pole bean is a legume that twines up the stalk of the corn plant. It is rich in protein, which is why it is a marvelous nutritional complement to corn and contributes to the fixing of nitrogen, the very chemical element that corn extracts in largest quantities for its own growth.

Wild grasses also grow in the spaces between corn plants, selected and promoted by ancient cultivators and modern peasants alike. Some of these are edible, such as pigweed. Others, on the contrary, are only tolerated and fostered because of their nitrogen-fixing properties. It is worth noting that corn also certainly possesses that property, unique among cereals, even though the rate of nitrogen fixing is lower than the rate of nitrogen extraction. Mixed farming and the wild grasses that are tolerated by cultivators allow soil nu-

Botanical Economy

trients to be replaced and makes the continuous use of the soil possible. The diverse plant community growing up in corn fields, to the horror of many scientific agronomists, serves much the same function as crop rotation and fallow elsewhere. Sowing corn among other crops, intercropping, and corn's ability to coexist with selected wild plants fulfill two functions. First, this approach increases total soil yields measured in annual cycles if we include the yield of the additional crops. Second, it elevates productivity in the long run by permitting the continual rotation of the same crop on the same plot. Intercropping also allows producers access to dietary complements with little additional labor input and little additional cost.

Since the 1930s, especially in countries with well-developed commercial agriculture, yields for corn and for other cereals increased appreciably with the introduction of hybrid seed. Corn was the first cereal to use hybrid seed and witness its widespread commercial use, taking advantage of the monoecious character of the corn plant. That is, corn has separate masculine and feminine flowers on the same plant: the masculine flowers on the ear or tassels that crown the stalk, the feminine flowers on the silks adhering to the stalk. The separation of the flowers made artificial pollination possible and economical. Such a process was much more difficult in other cereals in which the masculine and feminine flowers were together. The success of hybrid seed was so great that in some countries freely pollinated varieties, stable in their productive characteristics, were no longer cultivated. Much effort is being directed toward the pressing task of harvesting and preserving stable corn varieties and races. The creation of the new hybrids depends on these collections or seed banks. Nevertheless, these efforts are uneven: the majority of genetic banks are found in wealthy nations and many of these are in the hands of private organizations concerned with commercial applications. Thus, the cultural achievement of people worldwide who domesticated and developed the cultivated plants and endowed us with a genetic fortune is being converted into a private patrimony.

Let us return to our guide, Doctor Juan de Cárdenas, and let his voice from the past explain yet other virtues of our protagonist:

Third, for the ease and swiftness with which it is mixed and seasoned, since we see and we know that with wheat it is necessary to thresh it, grind it, sift it, knead it and then let it rise and cook it and even leave it from one day to the next, in order to improve it and be able to eat it with no harm done, adding salt, yeast, warm water and putting it in a very moderate and fitting oven, for an amount of time determined by the quantity of the dough.

Corn needs nothing of this: it is ground on a stone and mixed on this same stone and bread made, without adding more salt, yeast, or leavening, nor any other spices other than a little cold water, and it is toasted or cooked in the act in a dish or terra cotta comal, and is eaten warm as the most flavorful treat in the world, and made so quickly that one can be seated at the table and the bread has yet to be made, that I don't know what more good things can be said about bread, that besides being so good and such good nourishment, it is so easy and cheap to season. Fourth, for the brevity and swiftness with which before, as they say, once born it begins to nourish man, because since the time the very young ear begins to form, nestled inside the small pouch of a leaf that the Indians call *gilote*, and after forming kernels, being as they say green and later hulled, always serves to nourish and as a mouthwatering treat, as is the green corn after being roasted or cooked, so that luckily in this it also has an advantage over all other grains, for none of them can be used before maturing and ripening, and corn is even before it fully forms and can be called corn. (Cárdenas, 1988: 171–72)

Doctor Cárdenas implicitly emphasizes the fact that corn is a cereal, a form of grain that is the most efficient, compact, and natural way to store a seasonally produced food. Legumes also enjoy this trait to a certain degree, but few other cultivated plants can claim this characteristic. Perhaps this explains the predominance of grains as basic foodstuffs, the staple food of the human race, providing two-thirds of total human nutrition. Corn has other virtues as well. Green corn, one of its most flavorful forms, matures extremely early. A thinly veiled craving on the part of the good Doctor Cárdenas and this author is detectable behind the otherwise solid case they make for the virtues of green corn. Corn is also extremely simple to prepare. These properties translate into tremendous advantages for corn producers in terms of self-sufficiency. Corn can be consumed directly by those who grow it. Corn is easily prepared and does not require complementary or complex equipment in order to be consumed. That is, corn implies a certain degree of autonomy and independence for producers when compared to the centralized, costly, and complex services that other grains require in order to be transformed into food. In times of food shortages, corn has a short turn-around time and can be stored easily for long periods of time for use in the event of adversity. Everything can be done by a peasant family at home, using their own resources and without resorting to large-scale public and private services, whether of a social, economic, or technical nature. In times of crisis, when social services collapse or cannot

Botanical Economy

effectively carry out their functions, corn's importance becomes self-evident. Recourse to corn is the last line of defense for security, for hope, for the retreat of the lesser units of society in order to defend their very existence. When corn is scarce there is hunger, malnutrition, and rampant disease, and the very dissolution of those societies that depend on corn in order to survive.

The autonomy intrinsic to corn also made that plant an apt instrument for new settlement, for the occupation of new lands, for the opening up of human frontiers. The aforementioned characteristics became crucial in marginal areas, where great distances separated settlers and population centers where complex services were concentrated. Corn was the food of choice for settlers in America and elsewhere in the world. Above all, in a good part of the world corn was the one enduring staple of peasant classes and societies. Corn represented a way of life and an organization of production that tolerated exploitation and dispossession well. Corn, however, never necessarily implied or required such burdens.

Once more we return to Juan de Cárdenas in the sixteenth century: "Fifth, one can take pride in the fact that there is no part of the plant that is not used: the stalk is used after it is dried, for large, very elaborate images are made from them gathering the stalks in a bunch, and they are much better than wood, the juice from these stalks makes delicious black honey, the leaf is very good feed for horses; even a spike that this plant sprouts at the crown, called *mihaui* by the Indians, also is used, for the Indians make bread from it" (Cárdenas, 1988: 172). All these uses have been preserved and still are practiced by Mexican peasants—including the religious images made from corn stalks, crafted in Pátzcuaro even today.

Our chronicler fails to mention many other traditional uses for corn still practiced by Mexican peasants. The bractea, the leaves that protect the ear, are used as casings for some dishes, such as tamales of steamed corn dough. They also are used to wrap products being taken to market, including copal, an aromatic resin, and loaves of unrefined sugar. The husks are used to fashion artisanal toy figures. These dolls still are seen in the markets in Mexico and elsewhere in Latin America. The dried cornstalk serves as construction material for fences and for houses known as tlazole, in which the stalk is used to construct a wall or to frame a wall to be plastered with mud. Recently, there has been some success in the manufacture of paper made from the stalk and other parts of the corn plant, but the cost is still too high to make such a process competitive. Corncobs had a sanitary use as a precursor and substitute for toilet paper, although in the interest of discretion this is mentioned only infrequently in the literature. Cobs also serve to hull corn and, especially in the United States, to make pipes. In times of extreme want they are used as fuel, although they are not a very efficient heat source. The leaves of the corn plant remain excellent forage that serves as feed for draft animals. The colorful kernels can be used in mosaics, as seen in folk art pictures and paintings. There are several medicinal uses for different parts of the corn plant, something Doctor Cárdenas writes about at length. These include steeping corn silks, the pistils of the feminine flower, in hot water to use as a diuretic and tranquilizer. Corn dough is an effective poultice for wounds. This makes perfect sense in light of a subsequent revelation that corn dough is a favorable medium for the growth of the same type of fungus that is used in penicillin. Even diseased corn has a use in Mexico. While everywhere else *cuitlacoche*, or corn smut, condemns corn fields to destruction, in Mexico the blackened kernels are transformed into a most delectable and highly prized dish. All the corn stubble that remains in harvested fields serves as forage for grazing livestock or as green fertilizer once it is plowed under. Many other uses could be added to this list. This partial inventory does not even begin to illustrate to what extent American cultures came to exploit every part of the plant. These cultures preserved and actually prevented any loss of energy stored by the plant or any materials potentially derived from it.

Besides its traditional uses, the corn plant has other novel uses as well. The most important of these is growing corn for fodder. To this end, the green plant is cut whole just above the root and deposited in fermentation silos. There it becomes feed that can be stored for long periods of time. Corn is the only ensiled plant that has the capacity to fully satisfy the caloric needs for fattening large livestock. By virtue of corn's high caloric content, the simple addition of a protein supplement makes for a complete nutrient. Corn has been introduced for this purpose to regions where the kernel cannot mature due to short summers, as in England and the Scandinavian countries. There, corn becomes the most productive of the forage crops and the most efficient resource for the production of biomass per acre. Corn's energy efficiency has led wealthy northern European countries to take a closer look at the economic viability of using corn as an alternative energy source. The technology already exists to produce high energy fuel through the transformation of vegetable mass into ethyl alcohol.

The technology for transforming corn into alcohol is well established and quite respectable. Chicha, an alcoholic beverage made from fermented corn, played a central role in the ceremonial and social life of Andean peoples before contact with the Old World (Cutler and Cárdenas, 1947). Distillation, introduced after contact, improved that process. We need look no further than the potent corn brandies hailing from northeastern Mexico, or the good corn whiskey or bourbon. The first commercial distillery for corn whiskey was actually the brainchild of a nineteenth-century Protestant preacher. On several occasions, such as in the United States during World War II, ethyl alcohol made from corn and other grains was produced on an industrial scale. This, of course, was not for human consumption. The assessment of fuels derived from the fermentation and distillation of plant matter changed once oil prices began to rise in 1973. The technology for using corn-based fuels to keep a motorized and electrified world running were no long a dream or a luxury, especially when what once appeared as only some picturesque episodes in the history of corn apparently signaled an apparently irreversible long-term tendency. Alcohol extracted from sugar cane is commonly used for fuel in Brazil. In the United States, the world's largest corn producer, Congress approved enormous subsidies for the production of ethyl alcohol from corn. Gasohol is produced from mixing ethanol mixed with petroleum-based gasoline in a proportion of 10 to 15 percent. This fuel burns cleaner than gasoline and was intended to lessen U.S. dependence on imported crude oil.

Producing energy from corn could become the centerpiece of U.S. strategic policy. This would have many implications in the international arena. Such a project has not been successful thus far due to the falling price of oil. The consumption of oil worldwide likewise fell in response to economic recession and measures to encourage fuel economy. Beyond that, corn already occupied a central place in U.S. strategic policy. A full 30 percent of the U.S. harvest, around 66 million tons, was earmarked for export in the late twentieth century. That volume represented two-thirds of the corn that traded on the international market and constituted one of the most important sources of earnings from foreign trade for the United States. Many countries, Mexico among them, depend directly or indirectly on that corn as their major source of food. Food power is strategic power, and acts as one of the key factors in international relations according to a former U.S. secretary of agriculture (Butz, 1982).

The final virtue for which corn merits universal recognition, according to Doctor Cárdenas, is that "corn enjoys an advantage over all grains in terms of the many and different things made and composed of it, because not only are several types of bread made from corn, but eight to ten types of *atole* and starch, couscous, rice and other healthy and useful kinds of staple foods are made, since in order not to go on too long I say that one cannot desire anything more in the world than to see how without kneading it, one can eat corn, by itself merely toasted, like he who toasts chickpeas, or cooked or ground into powder, or dissolved in water and drunk, so that if one comes to think about it, one cannot wish for a staple that in and of itself has everything that precious grain of corn does" (Cárdenas, 1988: 172).

Once again, Doctor Cárdenas stops short even though he points out the obvious: the enormous variety of foodstuffs that Native Americans prepared in the sixteenth century using corn. José de Acosta, writing in the same period, sums it up when he writes: "So corn is utilized in the Indies for beasts and for men, for bread and for wine and for oil" (1940: 171). The enormous variety of corn dishes has not dwindled with the passage of time. On the contrary, it has grown. *Recetario mexicano del maíz*, published by the Museo de Culturas Populares de Mexico in 1982, includes more than 600 recipes in nine chapters, which despite its breadth does not claim to be an exhaustive inventory. If other recipe collections from elsewhere in America and around the globe were added to that compilation, the result would be one of encyclopedic proportions that would have given a great deal of satisfaction to our chronicler.

According to the Food and Agriculture Organization of the United Nations (FAO), there are eighteen countries worldwide that directly consume corn as their principal foodstuff. Dietary dependence on corn is extreme for the more than 200 million inhabitants of twelve Latin American and six African nations. In almost all cases, corn provides upward of two-thirds of all nutritional components in the diets of the inhabitants of these countries. Such is the case of Mexico. Many millions also depend on corn as their principal foodstuff in countries such as China or India, that as a general rule depend more on other cereals. The hundreds of millions of persons in the world who routinely consume corn but that in statistical terms do not depend overly much on it also must be taken into account. One can figure, conservatively, that a fourth of the world's population consumes corn in a direct form, as a matter of course, and depends on it to an important degree for their subsistence. This is the framework in which popular culinary tradition should be placed, one in which corn is eaten in every way imaginable and in some ways that even go beyond the imagination. The direct consumption of corn is one of the pillars of world nutrition.

In spite of the outpouring of popular culinary imagination, that is not the most significant change in the use of corn. Today, probably more than half of the world grain harvest is not earmarked for direct human consumption at all. Rather, corn is increasingly used for animal feed, and in turn those animals produce milk, eggs, and meats of all kinds. This is a form of indirect corn consumption. A good portion of the nutritional component is lost during that transformation, due to the low efficiency of converting calories and proteins of vegetable origin into foods of animal origin. On average, ten units of vegetable protein are needed in order to produce one unit of animal protein. On the other hand, protein of animal origin is of higher quality, although the price of the final product is higher in order to more than compensate for this.

There has been a more or less recent shift in the diets of wealthy, industrialized countries and of privileged sectors in poor countries from plant-based diets to diets depending mainly on foodstuffs of animal origin. Such a transformation, even for the benefit of the high caloric content of meat-based diets, cannot be explained without reference to corn and its expanding production the world over. It does not seem likely that this tendency toward meat-based diets, an expression of unjust and unequal growth, can go on indefinitely. Neither does it seem reasonable that it should unfold this way in the face of malnutrition and famine that affect such a great part of humanity. It has been calculated that in the United States alone the nutrients that are lost in the transformation of plant foods into foods of animal origin would be enough to entirely and satisfactorily feed 70 million human beings a year (Ebeling, 1979: 200–208). Another widely held estimate maintains that the grain utilized worldwide to feed livestock would be enough to satisfactorily feed the population of China and India combined (Kahn, 1984). In that transformation of plant nutrients into meat, not only does the taste for certain types of foods come into play, but also, and especially, enormous economic interests that are generated and configured all along the food chain.

In the second half of the nineteenth century, the first modern industrial corn-refining plants appeared. These plants transformed and utilized the different parts of the grain. The technology that permitted this process, wet milling, was based on the same principle and was in fact very similar to the thousand-year-old pre-Hispanic method for grinding corn. Cooking corn with lime before grinding was an essential step in the elaboration of tortillas and the extraction of starch. In the late twentieth century, the corn-refining industry constituted an enormous network producing hundreds of finished products. The corn-refining industry consumed between 5 and 10 percent of the world corn harvest. In the United States alone, the capital invested in the corn-refining industry approached $4 billion and employed around 20,000 workers. The products of that industry were essential to a twentieth-century world: industrialized foods, processed and transformed for long-term storage and easy preparation. Nutritional needs of highly industrialized and urbanized countries and a significant and growing proportion of nourishment in urban areas of poor countries was based on such processed foods. A typical symptom of this new way of eating was the supermarket. In the United States, according to a sample, a good supermarket offered some ten thousand products, including several different brands and varieties of the same product. Of these, one-fourth contained products derived from corn, almost all produced by the refining industry (Corn Refiners Association, n.d.).

Starch is the most basic commodity produced by the refining industry, and is an essential carbohydrate for human caloric intake and good nutrition. Cornstarch is no novelty, since it is mentioned by Doctor Cárdenas and it surely predated him. What are novelties are the derivatives produced by the corn-refining industry that, unlike starch, are water soluble. These are basically five: dextrines, for industrial use as adhesives; syrups, sweeteners, and artificial colors very common in industrialized foods; new syrups with a high fructose content and with much greater sweetening power with fewer calories; maltodextrines, an edible derivative of the dextrines that appears in all instant foods; and glucose or dextrose, corn sugar that is identical to that which flows through the blood of mammals, including human beings, which, in the final analysis, keeps us alive. Hundreds of specific products are derived from those starch derivatives. These products appear in almost all industrially produced foods. They also have specific industrial uses in the production of explosives, in drilling for oil, and many other uses as well. Additionally, nonsaturated corn oil for domestic consumption is extracted from the germ of the corn kernel. Other corn by-products serve as feed for animals, as culture mediums for the production of antibiotics, or as raw material for the manufacture of vitamins (Corn Refiners Association, n.d.; Walden, 1966). The wide array of uses for corn by-products produced by the corn-refining industry has implications above and beyond the account provided here. Once again, corn has a special distinction with respect to the other cereals: corn's full and complete incorporation into the industrial era and into modern capitalism.

The good Dr. Cárdenas would be more than satisfied with the universal recognition of corn. It is an exceptional and surprising plant. Its characteristics radically distinguish it from other cereals. For many, corn is a fluke, even a monstrosity. For others, it is the most evolved member of the plant kingdom and occupies a position comparable to that of human beings in the animal kingdom. The analogy is appropriate, because corn is clearly the offspring of humans, a gradual and impressive product of human invention, much closer to them, in a certain sense, than to any other living beings. Old World cereals have wild varieties that are propagated in nature. Corn does not exist in a wild state. Plants that are corn's possible progenitors—just as wild cereal varieties are for Old World cereals—are substantively different from the corn plant. Corn does not exist in a wild state because the plant cannot reproduce without human intervention. The remarkable ear of corn concentrates seeds in an orderly fashion and swathes them in a protective wrapping, all so that people can reap the benefit. The husk prevents corn from naturally dispersing its seed in order to reproduce on its own. The ears, with hundreds of compact kernels that appear simultaneously and compete to the point of destroying each other, do not produce viable plants when left to the vagaries of nature. Without human labor to separate and disperse seed, corn would disappear in only a short amount of time. People and corn depend upon each other in order to subsist and survive as a species. They are members of the same close-knit club, almost a clan. The millions who have domesticated these plants on the new continent have come into a valuable inheritance. In the course of their collective labor, they have accumulated and at the same time diversified genetic materials and knowledge and invented corn, a human offspring, our plant kin.

: 3 :
A BASTARD'S TALE

Corn is original to American soil and American civilization. Solid data and well-established theories have even made it possible to determine when and where the plant first was domesticated, wrested from its wild state only to become dependent on human caretakers. Our knowledge of those events is the result of modern scientific research.

Corn was subject to much more attention, passion, and debate on the part of the scientific community than any other plant over the last one hundred years. The purpose of this scientific research was to discover corn's origins. The fascination with the corn plant was motivated by the traits that distinguished corn not only from other indigenous plants but from other cultivated plants as well. There was a mystery there, a mystery of a botanical and historical nature that the scientific community still has not resolved completely. Scholarly interest and the passionate arguments revolving around corn's origin predate this more contemporary research. Ever since the sixteenth century there has been an ideological component surrounding the debate over corn's origin. This debate revolved around what was believed to be an inherent inferiority of American nature and American civilization as compared to the Old World. At the heart of the matter were the legitimacy of the domination of the Old World and, later, that of the temperate latitudes over the Tropics. The problem of corn's origin remains superimposed on the debate about the march of civilization and the advancement of humanity. To this day, the apparent neutrality of scientific language masks a considerable number of inherited ideological prejudices.

The nineteenth-century Swiss naturalist Alphonse de Candolle is considered to be the founder of the scientific method used to establish the origin of cultivated plants. According to his methodology, four types of analytical evidence are pertinent for this task: botanical, philological or linguistic, historical, and archaeological. On the basis of data available in the nineteenth century, de Candolle voices his position favoring the American origin of corn and confirms that the plant was unknown in the Old World until after Columbian contact. This hypothesis continues to be valid today, now backed up by much more data and research.

Ancestral wild corn is extinct. This has rendered botanical evidence confusing, even though there is reliable data on the geographical distribution of different native and cultivated varieties of the corn plant. Corn's closest indigenous relatives, teosinte or *Euchlaena mexicana* and *Tripsacum*, grow exclusively in America and provide a piece of evidence in this botanical puzzle. Nikolai Vavilov (1951) and his colleagues greatly contributed to the analysis of the distribution of cultivated corn varieties, staunchly promoting the hypothesis of the American origin of corn. Beginning in the late 1930s, advances in genetics and cell biology began to influence analysis of botanical evidence and ultimately helped to tip the scales overwhelmingly in favor of the American origin of the plant. Such a conclusion was based not only on the accumulation of positive evidence pointing in this direction, but also on the absence of countervailing evidence outside the New World (Mangelsdorf, 1974). Today there is almost universal agreement on the American origin of corn. Discordant voices periodically echo in favor of the Asiatic origin of corn and serve to revive the debate.

The oldest and most energetic debates revolve around the linguistic evidence. Many of the positions that reject the American origin of corn or that date the migration of the plant long before Columbian contact in the fifteenth century are based on the name for corn in other parts of the world. The brief summary of this passionate dispute that follows below cannot do justice to the erudition and imagination displayed by the participants. Neither can it reflect the imagination or unyielding determination that frequently arises in the course of that conflict. The subject has been the basis for an academic battleground in which the weapons of choice are textual citations and their interpretation.

Corn has a name all its own in almost all New World languages. This name is particular to corn and corn does not share its name with any other plant. In many languages the individual parts of the plant also have their own names that cannot be applied to the equivalent parts of other plants. The stems of those names are found in the linguistic trunks of New World languages. The American names for corn clearly suggest a long historical encounter with the plant, its products, and its uses.

In the Old World, on the contrary, the word for corn in most European languages was not unique to that plant. In some instances corn was designated by terms borrowed from another language, as is the case with the adoption of the term *maíz*—originally from Taino—in the Spanish language. In others, the name of another plant or plant product familiar in the Old World was used to refer to corn. Christopher Columbus called corn *mijo* in Spanish, or "millet" in English, after the Old World cereal. This was the case for the Portuguese as well, who also referred to corn as *milho*, or "millet." In British English, the word "corn" became a generic term used to refer to grain in general, but especially to wheat. It was not uncommon for a generic name or the name of another grain entirely to be added on as a qualifier in order to distinguish and designate corn as a new addition to Old World agricultural tradition.

Such an array of naming practices have caused a good deal of confusion and misinterpretation about the many names used for corn in the Old World. Geographic qualifiers often were added to an established term in order to distinguish corn from other plants. In many African languages, the name for corn means Egyptian grain or Egyptian sorghum. In Egypt, corn is known as Syrian or Turkish grain. In North Africa and in India, corn is referred to as grain or wheat from Mecca. In France and in Spain, corn is known by several names that contradict each other: Indian wheat, Turkish grain, Spanish wheat. In other parts of Africa corn is known as white man's grain and at times as Portuguese grain. It is also known simply as foreign grain. Of all the geographic qualifiers, that of "Turkish," commonly used all over Europe and North Africa to denote corn, stands out. In this case, the qualifier "Turkish" seems to imply something alien or foreign, something especially different and with origins in far-off reaches (Messedaglia, 1927: chap. 5). Thus, many of the Old World names for corn, with their geographic qualifiers, coincide in attributing an exotic, foreign origin to that plant.

Beginning in the sixteenth century, when corn first came to be known by the new names contrived for it in the Old World, many authorities followed the lead of these geographic qualifiers in order to verify corn's origins, without bothering to locate credible evidence to support their claims. In Turkey, for example, the most common name for corn—*kukuruz*—was taken from Russian, and there was no proof of its cultivation prior to Columbian contact. Nevertheless, the tradition of looking for corn's origin on the basis of the many names it has been known by gave rise to a voluminous body of work. Established experts were cited either to confirm or deny the American origin of corn or to deduce the convoluted migratory routes corn may have followed. This is part of the price for allowing academic discourse to hold sway over the facts.

The tendency to name corn after existing Old World grains was responsible for tremendous confusion in the linguistic data. Corn was known as sorghum,

A Bastard's Tale

wheat, millet, panic grass, and other cereal names, to which a differentiating qualifier was added. In some cases, especially all those in which corn took the place of an established cereal, the qualifier was dropped and corn simply came to be known by the name of the other grain. Luigi Messedaglia (1927) dubbed these dangerous homonyms. Some authorities mistakenly thought they actually had found references to corn in ancient Old World texts, such as the Bible, Herodotus, and the classic Chinese or Indian texts, among others, that referred to cereals whose name corn happened to inherit. These findings supported those who proclaimed that corn originated in the Old World. Their claims were in some cases backed up by the brazen falsification of documents, like that which came to be known as "The Corn Letter" (Riant, 1877). When the evidence in favor of the American origin of corn became overwhelming in the twentieth century, investigators such as M. A. W. Jeffreys (1971) once again resorted to these dangerous homonyms to prop up old arguments maintaining that corn's appearance in the Old World predated Columbian contact. There was no evidence to support such a theory and it remained purely speculative, provoking angry academic exchanges.

The third type of evidence, the historical or documentary evidence, offers no support for arguments alleging that corn was known and used in the Old World before Columbian contact. Surviving American documents from before contact are relatively few. Even these, whether they be codices, stelae, or quipus, have not been deciphered completely. Corn clearly figures as a central element in this sort of documentation, even though the precise or complete meaning may escape us. Pre-Columbian historical traditions compiled by American chroniclers do document corn's importance and antiquity. Corn appears in religious pantheons. It plays a conspicuous role in the explanations and myths surrounding the origin of life, of creation, and of civilization. Corn also appears in historical registers and in fiscal or tributary documents. Corn is a central piece of the history of American peoples before and after contact, a history that in good part has yet to be written, to be organized, to be understood.

It was not until the late 1940s that archaeological evidence supplemented by botanical knowledge came to resolve the mystery of corn's origin. In 1948, excavations carried out in Tamaulipas, Mexico, by Richard McNeish and others working in New Mexico uncovered many corn remains, including ears the size of one's little finger, that were thousands of years old. They determined that given the age and the diminutive size of the remains, this had to have been domesticated corn that was the product of practicing agriculture. Other excavations in northern Mexico and the southwestern United States offered up similar results, demonstrating that previously domesticated corn had been introduced from outside the region. McNeish later undertook excavations in Central America, Chiapas, and Oaxaca in southern Mexico, although without finding any older remains. In the meantime, in excavations for the foundation of the Torre Latinoamericana in the heart of Mexico City, workers found pollen belonging either to corn or to one of its wild ancestors, still a matter of some dispute. The pollen was some eighty thousand years old, much older than human settlement of the New World. The combination of those findings clearly flagged south central Mexico as the region in which ancestral wild corn was transformed into domesticated corn.

In 1960 McNeish and his colleagues began their excavations in the Valley of Tehuacán, in the Mexican state of Puebla. They were only able to partially reconstruct the process of the domestication of corn and of other plants. The sites of the oldest human presence in the valley did not show any use of corn at all. McNeish and Paul Mangelsdorf considered the specimens they found to be the oldest existing remains of wild corn, while other experts argued that in all essential aspects these were remains of corn that had already been domesticated. Fossilized remains and other evidence of the use of cultivated plants in the valley after the appearance of corn multiplied: frijoles, chilis, sesame, sapodillas, squash, and avocados. Nevertheless, for over a thousand years, cultivated plants scarcely provided 10 percent of dietary nourishment, according to the estimates of the research team. Remaining essential nutrients came from what foods people gathered and small animals they hunted or trapped. Eventually, nourishment derived from agriculture grew to make up 30 percent of total nutrition and allowed for the establishment of the first permanent human settlements. Pottery and hybrid corn varieties followed. The rhythm of transformation accelerated. Irrigated corn cultivation even appeared. The Valley of Tehuacán became fully integrated with the great Mesoamerican civilizations, according to the sequence established by McNeish's team.

During that long process, corn found in the Valley of Tehuacán developed from diminutive ears to full-sized ears indistinguishable from modern varieties. Many of the changes documented in the sequence did not originate in the same area, which meant that the Valley of Tehuacán could not be considered strictly as the only region in which corn was domesticated. The domestication of corn and the development of specific varieties, according to McNeish, was a disparate and geographically fragmented process. It included the Valley of Tehuacán, together with other regions and other peoples of present-day south central Mexico, although he could not determine precisely the exact limits and boundaries of that process. Archaeological research

A Bastard's Tale

points to the domestication of corn as a historical and collective creation of ancient peoples who occupied that part of the vast American territory. It was a prolonged effort that required the interest and the passion of thousands of anonymous agricultural experimenters over dozens of generations. That was how this miraculous plant came to be.

The excavations in the valley of Tehuacán were probably one of the most complete series documenting the transformation from nomadic to sedentary life at that time, not only in America but in the world. Many questions about the history of corn remained unanswered, however, despite the historical depth of that sequence. Who might have been corn's wild ancestor, corn's very paternity, was one issue that remained unanswered. The established hypotheses on this issue successively converged, even while their respective supporters at times became separated into irreconcilable academic camps. One position, combatively headed by Mangelsdorf (1974), sustained that wild corn, today extinct, was corn's immediate predecessor. Another hypothesis very close to this, put forth by Paul Weatherwax (1954), postulated that corn and its closest relatives, especially teosinte, derived from a common ancestor: a grass today extinct. Opponents of those theories objected, arguing the inexistence of that ancestor as either a living plant or as a fossil. Mangelsdorf counterattacked, claiming that the wild variety disappeared after crossbreeding with the domesticated variety, which became dominant. In the meantime, the fossil evidence was excavated: pollen in the foundations of the Torre Latinoamericana and the miniature ear uncovered in Tehuacán. These permitted a plausible reconstruction of corn's wild forerunner.

A growing number of researchers favored recognizing teosinte itself as corn's immediate wild ancestor. This wild plant was ubiquitous in some regions of Mexico and Central America. It crossed freely and easily with corn. It had annual and perennial varieties. It produced hard edible kernels much smaller than those seen in modern corn varieties. It had no ears, and was able to disperse seed without human intervention. The plant's name presented some difficulties. What was considered to be its common name, teosinte or *teosintle*, was not widely used in the scientific bibliography. The plant name probably had been invented or spread by some researcher, borrowing from a little used name common to an unidentified area. The collector who went around asking for teosinte was likely to be disappointed. Teosinte, Nahuatl for "corn of the gods," did not appear in the ancient dictionaries of that language. The plant that we recognize today as teosinte probably had other names that were more widespread in pre-Hispanic times: *cocopi* or *cencocopi* and perhaps *acecentli* or *acicintli*. In the Valley of Mexico in the late twentieth century, teosinte is known as *acés* or *acís*, words clearly related to the ancient name. Thus, it may be that teosinte was not the common or popular name of the plant at all, and that the use of that name perhaps was quite limited.

The scientific name of teosinte was not unequivocal either. Teosinte had been known as *Euchlaena mexicana* ever since it was definitively identified in the nineteenth century. Later, the plant was classified within the tribe may-deae. Corn belonged to this same tribe. In 1942, Mangelsdorf and others proposed that the annual varieties be called *Zea mexicana*, reserving the name *Zea perennis* for the only perennial variety known at that time. Subsequently, researchers discovered a second perennial variety that they called *diploperennis*. The new names categorized corn and teosinte as species of the same genus (Mangelsdorf, 1974: chap. 3). Later still, Hugh Iltis proposed that teosinte be classified as a separate corn variety and called *Zea mayz* subspecies *mexicana*. Factions that were proponents of the original nomenclature did not accept the new classification.

This hypothetical school that contended that teosinte was corn's predecessor also resolved the paradox of the total disappearance of its wild ancestor. George Beadle (1982), one of the most outstanding proponents of this theory, showed that the grains of pollen found in Mexico City in strata older than human settlement corresponded to *teosinte tetraploid*, and not to wild corn. Even so, the origin of the difference between teosinte and corn still needed to be explained, which meant explaining the development of the ear. Iltis (1983) put forward the theory of a sudden mutation that he called catastrophic sexual transmutation, since it would impede natural reproduction and dissemination of any new plants. Such abrupt change would explain the lack of evidence of a gradual transformation. In that mutation a monstrosity arose, the ear, which by virtue of its position on the plant captured nutrients in a concentrated form and could contain up to a thousand kernels, while teosinte produced between eight and ten kernels, just like the wild cereals of the Old World. Iltis's theory on the origin of corn resolved most of the paradoxes that had arisen up until that time. Just as with all previous theories on corn's origin, however, Iltis's was not universally recognized, even as it increasingly won adherents and gained legitimacy.

Artificial selection through human intervention was the only thing that could explain corn's persistence and development to the point of dominating its wild predecessors and closest relatives. Corn in its natural state was incapable of dispersing its own seed, thus spelling its own doom. Native cultivators played an active role in collecting seed, raising seedlings, and nurturing the young plants. Only by a great stretch of the imagination could the human

A Bastard's Tale

intervention that propagated corn be explained as a casual coincidence or fate. A relation of constant practice and observation, of accumulation of knowledge about nature by ancestral peoples, was a much more likely scenario. This knowledge and the actions derived from it, such as intentional dispersal of the kernels of that peculiar plant and, later, its cultivation and hybridization in order to adapt and improve it, constituted a widespread and collective heritage, as suggested by the pattern established at the excavations in the Valley of Tehuacán. Thus, it was the common legacy of native peoples' knowledge and means to cope with nature and society—their culture—that rescued corn from suicide. It was culture that was responsible for pampering, nurturing, and improving corn.

The cultures of indigenous American peoples preserved the memory and possibly the history of the domestication of corn in their mythic literature. In the Aztec creation myth, men were created five times, each time more evolved and perfect; *cencocopi* and the *acicintli*—teosinte—were depicted as the principal nourishment of men in the two last unsuccessful creations. However, when man nourished himself with corn in the fifth creation, the world lasted until the time of the chroniclers (*Códice Chimalpopoca*, 1975). In the Mayan narrative *Popol Vuh* (1947), several attempts to create man had failed. Finally, at long last, a stable, enduring method was developed to mold men from a mixture of the blood of gods and corn dough: "From yellow corn and white corn his flesh was made; from corn dough the arms and the legs of man were made. Only corn dough entered in the flesh of our fathers, the four men that were created." In other narratives Quetzalcóatl was the one who introduced man to corn. Quetzalcóatl was the deity endowed with civilizing qualities, and so it was fitting that it was he who gave man the building block of civilized existence.

The domestication of corn as a cultural achievement cannot be judged by the same criteria used to judge the accomplishments of other cultures. Reclaiming in-depth knowledge about the process of the domestication of corn and corn's forerunners remains crucial in order to reclaim a rich legacy of botanical knowledge and genetic engineering. That knowledge also must work toward changing the dominant perception of peasants and indigenous peoples who still practice that tradition and transform it in the course of living their daily lives.

It is now time to return to our protagonist, to the mystery surrounding corn's status as a bastard. While that lack of legitimacy persists, an imminent denouement is on the horizon. It is time to remember something that is obvious but regularly ignored: if corn's paternity is in doubt, corn's maternity is not. That maternity is known and must be reclaimed. It can be found in the aboriginal cultures of present-day central and southern Mexico. There, thousands of years ago, nature and knowledge came together in order to create corn, the material sustenance of life itself, and the subject of a long and contradictory history.

4

CORN IN CHINA: THE ADVENTURE
CONTINUES HALF A WORLD AWAY

Columbus and his men discovered corn on their first expedition, as duly noted in the explorer's journal. There was nothing dazzling or spectacular about this encounter. Corn was just one more novelty among the many that caught the attention of the disoriented expeditionaries, who still were convinced that they had arrived in the Far East. The first recorded mention of corn is in dispute. Many experts point to a journal entry made by Columbus on October 16, 1492, on the Isle of Hispaniola, today Haiti and the Dominican Republic (Mesa Bernal, 1995). Other experts contest that entry as vague and speculative. Instead, these authorities propose a better-documented event at a later date. On November 2, 1492, All Souls' Day, Rodrigo de Jerez and Luis Torres—a Jewish convert to Christianity who claimed to speak Hebrew, Chaldean, and a little Arabic that would enable him to communicate with the Orientals—were sent out to look for a large inland city on the island of Cuba. They returned three days later without having found the metropolis. Instead, they came laden with gifts bestowed by friendly natives. Among these were corn and tobacco, according to the journal entry of November 6 (Weatherwax, 1954: 28–30). This apparently trivial discrepancy over the date corn was first documented serves to illustrate the elusive nature of our main character, its penchant for leaving nothing more than a faint impression, always subject to debate and interpretation.

Seed corn probably arrived in the Old World with the explorers returning from the first expedition, and corn most likely began to be disseminated at that time. Surely when Columbus and his fellow expeditionaries returned to Spain from their second voyage in 1494 they brought corn with them, since Pedro Mártir de Anglería submitted white and black corn kernels along with the manuscript of the second volume of his *Primera década del Nuevo Mundo* to his sponsor at that time (Weatherwax, 1954: 32). While this may have marked the date of the initial migration, it was not the only one or, of course, the last. Beginning in 1494, the dissemination of seed began, and we can assume that its dispersal was continuous and uninterrupted in all directions. We will never know for sure from how many points of origin and by way of which routes corn embarked on its peregrinations abroad. Indisputable although fragmentary evidence establishes that corn was already well known all over the Old World—Europe, Asia, and Africa—and some islands in the Pacific and Atlantic well before the end of the sixteenth century. Half a century after contact between the New World and the Old, corn would reach the ends of the earth: China.

We do not know for certain the routes or the dates of this extremely widespread and early migration. Even less is known about those who undertook to transport the seed, plant it, and reproduce it, or their reasons for doing so. Their motives? Curiosity and envy, perhaps akin to those who even today will take a plant cutting from public and private gardens alike in an attempt to reproduce some coveted and exotic ornamental plant. "Won't you give me a snippet from your plant?" is a plea repeated a thousand times a day in Mexico and that promotes democratic sharing out of genetic plant material. Farmers and merchants may have been moved by self-interest and ambition in their search for a profitable novelty. Collectors may have been motivated by the irrepressible nature of scientific curiosity typified by experimental single-mindedness. It may have been a purely practical matter, one of ease and common sense. Corn satisfied basic needs better and cheaper than any other grain. Corn was also an effective instrument in colonizing exotic virgin lands that themselves were a means to satisfy new social imperatives. Corn served strategic state interests, what today we would call national security and dietary autonomy. Corn also promoted fiscal interests in the improved collection and disposition of tribute and taxes, necessarily supplemented by diplomacy and espionage. European scientists in the New World, assigned to learn about plants and their uses, served precisely such a function. The structural problems of societies, state politics, incidents, and accidents, the banal and the frivolous, all that and much more, must have been a part of corn's early and fast-paced migration. From that complex tapestry of the human condition we can scarcely salvage a few strands, at most some small remnants, that serve less to answer questions than to raise them. Nevertheless, we can verify, document, and perhaps better understand the effect of that migration in the long run.

Matthieu Bonafous, in his *Histoire naturelle, agricole et économique du maïs*, published in France in 1836, reproduced a surprising and fanciful Chinese woodcut depicting a corn plant crowned by an ear occupying the place of the tassels. That illustration formed part of a treatise on the plant written by Li Shih-Chen between 1552 and 1578, published posthumously in the early seventeenth century. For many years the illustration was thought to have dated from the mid-sixteenth century. At one time it was considered the oldest representation of corn in the Old World. The text that accompanied the woodcut noted that corn was an exotic plant, introduced from the West, and that its cultivation was extremely limited. Nevertheless, that early image served as fuel for extreme diffusionists who maintained that corn originated in the Old World or that its migration dated from before the time of Columbian contact. That debate fueled the search for data and the accumulation of information that permitted a limited reconstruction of corn's introduction into China.

The earliest reference to corn in China prior to that illustration was a book published in 1555 on the history of Gongxian, a district in the inland province of Hunan in central China. Gonzalez de Mendoza, an Augustine monk, published a history of the great and powerful Chinese kingdom in Spanish in 1585. In it, he stated that the Chinese cultivated the very same corn that was the principal food of the natives of Mexico and Peru. His information came primarily from Martín de Herrada, who visited the southern Fujian province, on the southeastern Chinese coast in 1577. Other histories of Yunnan province, in southwestern China, written in 1563 and 1574, also documented corn cultivation there. The 1574 publication specifically mentioned that corn was being grown in six prefectures and two departments, making this the most extended area of corn cultivation in China. The lack of spatial continuity of those three points that formed a triangle led Ping-ti Ho (1955) to conclude that corn probably came to China by way of two separate routes: a maritime route, which would explain its presence on China's east coast, and an overland route along a line stretching from India to Burma and, finally, to Yunnan. The same author supposed that corn's introduction should have preceded any written mention of that plant by at least one or two decades and that corn's overland migration had to have been some several years earlier than that. This would imply that corn took just about half a century after contact to make its way to China.

Portuguese merchants and navigators certainly played a role in corn's maritime route to China. The Portuguese first arrived in the southern port of Canton in 1516. Peanuts—another American plant—were cultivated in southeastern China sometime before 1530, and their presence there is undoubtedly due to the Portuguese. Peanut cultivation demonstrated the feasibility of such a maritime route for the migration of American plants. The overland route corn could have followed also was documented by contemporary Chinese historians, who attributed corn's introduction to the Western barbarians. Non-Chinese ethnic groups that inhabited the mountainous frontier areas were collectively referred to as barbarians. Western science inherited the prejudice and labeled those groups as tribal and, therefore, barbaric. One of the first Chinese names for corn was "wheat of the western barbarians." Later, the names "jade wheat"—referring to the color of the plant—or "imperial wheat" became widespread. The latter name would seem to have to do with the fact that the ethnic groups on the empire's western frontier typically paid their taxes in corn. There were massive pilgrimages of thousands of people to Peking in order to meet this obligation, with the usual unpleasant exchanges with bureaucrats in the capital. The journey to pay taxes would explain corn's early presence in China's northern provinces. Corn's trail was lost on China's western frontier, and the rest of the overland route could not be reconstructed with any certainty. One could not reliably depend on inferences and suggestions for credible information on this issue.

Edgar Anderson, working in the laboratory, and C. R. Stonor, working in the field, produced a monograph (1949) on corn among the hill people of Assam. Assam was a province in northeastern India on the overland route to China, bordered by Burma, Bangladesh, and Tibet. Unfortunately, the work was based on a botanical description and a limited synchronic ethnography lacking historical depth. Rice patties were rare in that region. It was corn that figured so prominently in the agriculture of that hilly countryside. Corn was cultivated at all altitudes, but was especially common in the higher reaches. Corn was a seasonal summer crop, cultivated during the rainy monsoon season. It adapted easily to the extensive systems of slash and burn agriculture, in which corn was intercropped among other plants. Corn was the most important food for inhabitants of that mountainous region, even though given a choice, rice was the overwhelming preference.

For the cultivators of Assam, whom the authors treated as members of a primitive tribe, corn's presence was age-old. Corn had been there always and there was no memory of its introduction from the outside world. From the perspective of the heartland of the Indian empire, the inhabitants of Assam were primitive, marginal, and backward people who were not considered to be true Hindus.

Anderson and Stonor succumbed to a very common pitfall when they characterized the inhabitants of Assam as backward and primitive peoples. This led these authors to assume that the culture necessarily dated from ancient times. From those limited facts and inferences about marginality and backwardness as historical stasis, the authors revisited hypotheses concerning the Asiatic origin of corn. Their work put more emphasis on the differences in corn varieties than on agriculture and the history of the peasantry of that region. The critiques from botanical circles were immediate and devastating. Criticism of the cultural prejudice and of the deficient information on the cultivators never materialized.

Perhaps this is the time, at the risk of sacrificing continuity, to comment on the common prejudice that supposes backward peoples are living relics of the past, walking archaeological artifacts. The first problem is the descriptive term itself: "backward" as a self-evident category for which rigorous criteria is unnecessary. It never has been a suitable descriptive term. That the imperial prejudices of China and India happened to have coincided in their estimation of the mountain dwellers in their peripheries is no sort of criterion. It is more likely a reflection of relations of power and of domination. In that area of the world, the peoples on the peripheries of the great empires are the first to adopt novelties and innovations, long before those might penetrate more centrally located regions. Corn's rapid and flowing migration suggests that there is no isolation of peripheral peoples, among whom information and innovation are seamlessly transmitted. The "backward" peoples are peoples with their own history, neither slower nor more conservative than other histories, simply different.

To the east of Assam is Burma, another missing link in the chain of information, even though the renowned anthropologist Edward R. Leach (1954) worked among the Kachins, neighbors of Assam. Neither agriculture nor history were subjects of much interest to that field worker, who was concerned primarily with social and political structure. Leach makes only the briefest mention of corn and its place in the slash-and-burn agriculture of those people in his work. To the west of Assam is the Indian subcontinent. Corn is everywhere in this vast territory, but is most important in the northern provinces, where the Himalayas descend into mountainous terrain. There corn is the staple of the peasant population (Laufer, 1906). A wide band of corn winds its way through highlands and the steep slopes of the southern foothills of the Himalayas, from China to Afghanistan, an area inhabited by mountain peoples, somewhat similar to the terrain where corn was domesticated in America.

I did not find the documentation on how corn made its way to that Asiatic corn belt very persuasive, although corn's introduction could be set with some degree of certainty at sometime during the first half of the sixteenth century. Corn could have arrived by sea, given that Portuguese sailors navigated a direct and established route between Brazil and Goa after 1500. They set up a permanent post on the western coast of India that lasted for 150 years (Merrill, 1954). Corn also could have arrived overland by way of the legendary caravan routes established long ago, the paths of spices and of silk, the route of Marco Polo, in which Europeans did not directly intervene. Both scenarios were likely, and perhaps others could be added as well.

The mobile and fast moving nature of our protagonist could create the impression that corn's appearance in China was triumphal and resounding. Nothing nearly so spectacular took place. Its entry was through the back door of the empire, the service door, if you will. Corn was not alone. The peanut, the sweet potato, and perhaps the chili pepper, all indigenous American plants, accompanied corn in its early exploits, and they were later joined by the potato. The peanut, introduced for cultivation in sandy-clay soils, was not suitable for developed and fertile agricultural regions. In the first century and a half after its arrival, it came to be the most widely accepted of the American plants and soon began to turn up at imperial banquets, although it retained its character as a complementary or secondary food. The sweet potato, with its much higher yields, was capable of displacing the millenarian taro. The sweet potato rapidly gained acceptance in the southeastern provinces, where it became the staple food of the poor. In 1594, the state used pamphlets to promote the sowing of sweet potatoes in order to mitigate the effects of a disastrous agricultural cycle. China became, slowly, the world's largest producer of sweet potatoes, a distinction it still held in the 1980s.

Corn was accepted early on in the southwestern provinces and in the less developed areas surrounding them, the mountains and foothills, where the more advanced agricultural practices used elsewhere in the empire were not feasible. With few exceptions, corn came to China as a food for poor and marginal populations, for those ethnic groups that were not really considered to be true Chinese (Ho, 1959). Since the sixteenth century, corn was eaten in many different forms in China. Where harvested early when it was still tender, corn was eaten as a fresh vegetable. Where harvested already mature, corn flour was elaborated and breads and thick corn gruel were prepared (Laufer, 1906). Corn also was consumed as a fermented liquor or beer and at times distilled in order to make whiskey.

Corn and other American food crops were and are summer crops in China, utilizing seasonal rainfall or simply dry cropped. They were cultivated on lands that were not suitable for irrigation and rice patties. The peanut used sandy soils not suitable for other crops, while sweet potatoes and corn were sown in lowlands with sheer slopes; corn was even planted in steep mountainous areas. Initially, sweet potatoes, peanuts, and corn were not alternatives for rice, the preferred grain. Rice probably had the highest agricultural yield in the world in China in the sixteenth and seventeenth centuries. The production of American plants was reserved for marginal areas, undesirable and peripheral lands with respect to irrigated agriculture of the great river valleys. In this marginal position, American plants were used as reserve crops.

According to Ho (1959), several factors came together in order for American plants, especially corn, to thrive. First, the surface area and resources for increasing the acreage in labor-intensive irrigated rice patties reached a point of maximum use and saturation. Ho set the saturation point for all of China at around 1850, even though in many regions and provinces it was reached much earlier. The effect of the law of diminishing returns set in at this point and accelerated sharply after that time. The increasingly complex and difficult construction and the maintenance of hydraulic works was a precipitate of this development. The sharp drop in the productivity of labor was yet another manifestation of the law of diminishing returns. The productivity of labor fell off so much that the traditionally high levels of rice yields were only barely maintained in these fields. Their fertility had been severely compromised by working them continuously over the course of several centuries. On the other hand, China's population grew rapidly after 1700 and the geographical distribution of that population changed. The population of China's northern provinces grew more rapidly, even though in absolute terms the south would maintain its demographic supremacy. There was no lineal relation of cause and effect between the gradual close of the frontiers for expansion of intensive agriculture and the acceleration in the rhythm of population growth. There was, however, an unresolved correlation between a series of events that ultimately compelled corn and other American plants to emerge from the marginal position they occupied in order to play an important role in the agricultural development of China.

Corn cultivation was quite limited until 1700. Even though corn had spread slowly outward from what was supposed to have been its nuclei of introduction in Yunnan, in the southwest, and in Fujian, in the southeast, corn had not been able to get a foothold in the agricultural heartland of China: the Yangzi River basin. There, in the early eighteenth century, the hill country was still virgin territory in spite of the high demographic density of that region. The numerous inhabitants of the great alluvial plains dedicated all of their efforts to irrigated agriculture. In the eighteenth century there was a powerful migratory current originating in the southeast to settle the mountainous lands and the hill country of the upper Yangzi. The highlands in the region had already been peopled using American plants. That human mobilization had two currents. One continued its movement northward, following the Han River basin and introducing corn cultivation into the provinces of Hubei, Hunan, and Shansi. The other was to the west, eventually reaching the province of Sichuan, where corn was already cultivated in mountainous plots, just as in Guizhou, to where it had spread from the southwest.

American plants were the instrument for the productive conquest of space in that process of domestic settlement. Indigenous American plants grew in the mountains and on hillsides and in the richer and more developed agricultural areas on unirrigated marginal lands. Corn stood out among the American plants and soon displaced summer and seasonal Old World cereals. Corn leant itself readily to any climate and had high yields. Corn was easy to store, transport, and convert into food, all of which favored its adoption as a staple food. Agricultural settlement along the domestic frontier brought about what Ho (1959) called the second Chinese agricultural revolution: the occupation of marginal unirrigated lands, an act that opened the agricultural frontier at a time when the growth of intensive irrigated agriculture had reached its natural limit. Such a technological sequence was just the opposite of what we were accustomed to seeing, from our perspective on the other side of the world. That is, we took it for granted that the use of intensive irrigated agriculture was a stage subsequent to dry farming, not prior to it. We tended to assume that the sequence of dry farming–irrigated agriculture was the logical fulfillment of some inexorable law of progress.

With the second agricultural revolution corn never lost momentum, growing at a faster rate than other cereals and basic foodstuffs. In the late 1980s corn placed third, after rice and wheat, among basic foodstuffs in China, and corn production represented 20 percent of the total production of cereals (Food and Agriculture Organization, 1983). Corn stood alone as the most important seasonal summer crop, inasmuch as wheat was a seasonally sown winter crop and rice was basically an irrigated crop. Corn yields were higher than those of wheat, but less than those of rice. China led in annual volume of corn production during that period, almost 72 million tons, and was second only to the United States. China was the largest corn producer among those countries that earmarked corn primarily for human consumption. These production figures represented no small accomplishment, given corn's inglorious arrival in China only 400 years before, slipping almost unnoticed through the imperial back door.

The second Chinese agricultural revolution made it possible to sustain dietary self-sufficiency over a long period of accelerated demographic growth for an absolute population of enormous magnitude. In 1700, China—excluding Tibet and Turkestan and Manchuria and Mongolia—had 150 million inhabitants, the same number as in 1600. By 1800, China's population reached 320 million inhabitants and by 1900, 450 million. In 1975 China's population was already 720 million, 835 million if the populations of Tibet, Turkestan, and Manchuria were included, which formed part of the People's Republic (McEvedy and Jones, 1978). By the 1980s, with China's population at more than a billion people, the country's most basic requirements for dietary self-sufficiency had been met.

The basic dietary self-sufficiency the second agricultural revolution brought about did not imply that hunger, in its acute or chronic state, was absent from Chinese history between 1850 and 1950. Besides other more restricted and lesser crises, there were at least four very serious widespread regional famines over those hundred years. The result of natural catastrophes and social convulsions, their victims were estimated at 22 million dead. Chronic hunger was severe in the first decades of the twentieth century. Richard H. Tawney cited a Chinese official who declared in 1931 that in recent years three million had died and four hundred thousand women and children had been sold due to lack of food (Tawney, 1932: 76–77). Nevertheless, those painful losses were not attributable to insufficient production, but rather to problems in distribution. Either it was impossible to get food to those areas where an agricultural catastrophe had taken place or, especially, the political system was incapable of attending to the dietary needs of its most populous and poorest classes. Continuous warfare and other processes that accompanied the disintegration of an authoritarian and centralized empire accelerated geographic fragmentation and social polarization. The ancient institution of public granaries slowly deteriorated and saw its final demise with the advent of the republic in 1911. The granaries had been an effective regulator of supply and guarantor of reserves for times of crisis. Their collapse had an important role in bringing the problems of distribution of food to a head (Wolf, 1969: 128). Tight control over systems for monetary and mercantile supply, marked by speculation, in a country rent by geographic and social inequality, apparently played a more important role in the explanation of hunger as a phenomenon than low absolute levels of production in and of itself.

The second Chinese agricultural revolution was carried out by peasant families on small, sometimes minuscule plots, especially from the U.S. perspective. During that process, corn cultivation definitively became dissociated from specific ethnic groups and instead became linked to a widespread agricultural tradition. A 1917 census counted 50 million units in this category. Each one of them had access to an average of a little over three and a half acres. Some 18 million units, 36 percent, held not even one and a quarter acres. In general, even that extremely modest figure did not reflect land owned and worked as a single parcel. Rather, the land was divided into many smaller plots, between five and forty, and those tended to be far flung. Such fragmentation was desirable from the point of view of peasant owner-operators, since it divided the different attributes of the soil among all producers. Such a complex agricultural model that carved an intricate mosaic out of the land prompted Tawney to suggest the metaphor of pigmy agriculture in the land of giants (1932: 38–45). That agricultural model was essentially a form of gardening in which each one of the plants received individual attention. Plentiful peasant labor was applied in a labor intensive fashion to care for a great variety of plants in order to make the best possible use of the limited amount of land.

Many complex factors influenced the proliferation and development of that model. One of these was the relative scarcity of cultivable land. Only a little more than 10 percent of the total land area was arable. It was densely populated by the agriculturists who farmed it. In the first half of the twentieth century at least three-fourths of the Chinese population was rural and worked in agriculture. There were no lands available for pasture or stock breeding that might compete with agriculture in China, with the exception of the northern provinces. Agricultural production did not tend toward forage crops. Not even corn was used for this purpose. Rather, corn was and continued to be a direct food for the population. Human labor was the principal source of energy in agriculture and draft animals supplemented this in only a few areas. The means and knowledge to better exploit the productivity of manual labor in China constituted one of the world's richest and most varied repertoires, a catalogue of monumental proportions that brought natural materials and human ability together (Hommel, 1969). The millenarian hydraulic tradition surrounding irrigation, the social axis and basis for the state, combined abundant labor with the division of land into small units to guarantee the land's most intense and efficient use. Rice was the preferred and most important staple food of Chinese civilization. Rice was grown on irrigated lands that produced extremely high yields, perhaps the highest in history for any cereal before the twentieth century.

Social factors contributed to the technical, demographic, and geographic factors outlined above. An age-old inheritance pattern divided assets equally between all descendants. That practice worked against the emergence of large landholders or a large landowning aristocracy, nonexistent classes in China. Any extensive properties that did arise became hopelessly fragmented in just a few generations. There was no egalitarian access to land. Rather, a complex network of mercantile and private relationships were an integral part of land tenure. Even though in the past the state and its institutions had been important landowners, in the twentieth century more than 90 percent of agricultural land was in private hands. Land belonged either to individuals or to clans, corporations of relatives that preferentially rented it to clan members at commercial rates. In the first quarter of the twentieth century, nearly half of all peasants owned the land that they worked. A third had access to land through sharecropping or leasing and another fifth supplemented insufficient property with sharecropping or leasing.

Rental or leasing arrangements were complex and varied, but in general such agreements created stability in access to land. In contrast, sharecroppers had to pay out a significant proportion of the value of their harvest, in cash or in kind. Estimates for the first half of the twentieth century placed the price of ground rent at around half the total value of production. Rents supported a landowning class that, while it neither possessed nor managed large holdings, controlled a great proportion of production and agricultural surplus. This class had other functions: they were usurers, agricultural middlemen, merchants, and, frequently, local government officials and dignitaries who administered the assets of their clans. Those "quadrilateral beings," as Cheng Han-seng called them (Wolf, 1969: 132), kept the majority of the peasants in poverty, on a strict subsistence level, and at times on the threshold of starvation. Poor peasants made up the vast majority of the population. The very same standards that maintained and exploited peasants also prevented an agricultural proletariat from developing. Figures in this respect tend to be quite misleading by virtue of the fact that owners or tenants were often patriarchal or extended families that consisted of several nuclear families of different generations. Despite this, calculations almost never put the proportion of peasants without direct access to land at above 15 percent of the total (Tawney, 1932; Wolf, 1969: 101–55; Moore, 1966: 162–227).

There has been much debate about China's millenarian stability and resistance to change. I am not the one to take on that debate, although I do have an opinion on what seems to me to be an exaggerated position. Chinese history is full of change, of innovation, of rupture and reconstruction. Like all histories, Chinese history may have its own structure and style, its specific limitations and restrictions, that can give the appearance of stasis to outsiders. The second Chinese agricultural revolution, in which corn and other American plants become widespread in agriculture and in the Chinese diet, takes place during a period of great unrest. This transformation emanates from instability and dispossession, from disorder, from the mobilization of peasants to survive and to change for the better.

Ho, above, located the principal cause of the second agricultural revolution in the saturation of the growth of irrigated agriculture in the great alluvial plains around 1850. The reasons for that saturation were complex, but could not be attributed to natural impediments or to technical limitations alone. During the Opium Wars, which took place between 1839 and 1842, British cannons forcibly opened Chinese ports to free trade, to the importation of cotton textiles and opiates cultivated by the English in India. Until that bellicose episode, trade between European powers and China had already been going on for three hundred years, albeit in a restricted and controlled form. The physical presence of Europeans on Chinese soil and their sphere of activity in the empire also were strictly limited. Other wars took place after the Opium Wars in order to consolidate free trade and the privileges of foreign powers in China: the Anglo-Franco War of 1860–61, the annexation of Vietnam by the French, the war with the Japanese in 1894–95, the Russo-Japanese war that was fought on Chinese soil between 1904 and 1905. Cynical speeches lauding the civilizing attributes of the West attempted to mask the naked economic motives behind these brutal foreign aggressions. Internal rebellions, lasting longer than the wars with foreign powers, aggravated the situation even further: the Taiping Rebellion between 1850 and 1865, the rebellion of the Nien from 1852 to 1868 (Wolf, 1969), and the Boxer Rebellion between 1898 and 1900, this last conflict brutally repressed by Western armies (Harrison, 1967).

The effect of that tumult was disastrous. The debilitated Chinese state disintegrated. Not only did it lose effective control of its own territory, economy, and populace, but it also was financially ruined by the punitive payment of indemnities to the victorious aggressor governments. One of the long-standing and essential functions of the state had been the construction, maintenance, and administration of hydraulic works. Without financial resources, effective control, or legitimacy to carry out that function, state administration of those works lapsed. The result was not only the stagnation of the hydraulic works, but their progressive deterioration, resulting in disastrous floods that claimed millions of lives. The disintegration of the state appeared to be one of the pivotal causes of the saturation of irrigated agriculture and, consequently, the second Chinese agricultural revolution. This could be proven indirectly by the fact that irrigated surface area doubled in China after 1949, with the triumph of the revolution and of the reconstitution of a hegemonic nation state.

Through violence, China became a satellite in the economic orbit of nineteenth-century Western powers. Such a status meant that China enjoyed neither development nor modernization. Rather, China experienced only inequality, deterioration, violence, and, finally, revolution. This is not the place to review the effects of the incorporation of China into the economy of foreign powers, but the consequences of this incorporation on the peasant population and the rural economy can be summarized using the work of Eric Wolf (1969: 127–28). In the first decades of the twentieth century, the disequilibrium between resources and the growing Chinese population produced almost unbearable agrarian tensions, aggravated by the neglect of hydraulic works. As centralized power grew weaker, regional and local warlords appropriated imperial taxes for their private purposes and imposed new and onerous tributary levies on the peasant population. The peasants received nothing in return for that exaction. Looting by private armies and the destruction of public granaries dismantled supply mechanisms and left the peasant population with neither resources nor reserves with which to surmount times of crisis. Surpluses generated by the rural producers and expropriated through rent, usury, and middlemen, age-old phenomena, were transferred from the countryside to new activities and economic zones. New phenomena that arose were protected and promoted by the incorporation of China as a satellite in the orbit of Western powers.

In that new context, usurious interest rates and rising rents intensified the absolute rate of extraction of surplus. The terms of exchange with other sectors of the economy deteriorated as well. What peasants bought and sold was priced out and paid in copper coin, by virtue of its small monetary value. Rarely were transactions large enough in monetary terms that they might merit pricing out in silver. What was certain was that the most prestigious and most widely circulated silver coin in China in the first decades of the twentieth century was the Mexican *peso*, that country's long-lost hard currency. Corn was not the only thing that departed Mexico for China. A traditional and century- old ratio of 1:2 was practiced for silver to copper. Silver had accumulated in China over several centuries of trade, and was now extracted to pay for imports and indemnities to Britain for those wars that effectively strong-armed China into free trade. Silver became scarce and the balance of silver to copper changed to 1:3. Agricultural commodities were devalued, since they were rendered in copper, while many other necessities, domestically produced or foreign imports, became more expensive because their prices were rendered in silver.

Under those conditions, the participation of peasants in commercial agriculture designed to produce money rather than consumer goods grew exponentially. The cultivation of commercial crops destined for the market, like tobacco and opium, grew at rapid rates, diverting land and labor from the production of foodstuffs. Also, relatively scarce staple foods entered the commercial sphere, with willing urban markets already in place. This especially was the case with rice, a perennial favorite. "Such is the value of rice as a means of obtaining cash, that many farmers eat their own rice only at special times of the year. They prefer in many cases to dispose of their entire crop with the exception of seed and invest part of their return in cheaper foodstuffs for their own consumption. Thus many Ch'uhsien rice growers eat maize as their staple" (Fried, 1953: 129).

Growing rural surpluses did not remain in the rural countryside or even in China itself, for that matter. They were transferred to foreign powers' spheres of economic influence and accumulated there. Peasants were the source of agricultural know-how and labor, yet they increasingly were threatened as they took part in the second agricultural revolution, settling marginal lands on the nation's domestic frontier. For many decades they accepted the destiny of peasants everywhere: unable to eat what they produced because it was prohibitively expensive. Thus they transformed corn and other American plants, previously foods for the poor, into essential resources for their very survival. They did even more than that: they carried out a revolution.

The second agricultural revolution was part of the same process that moved peasants to social revolution, a revolution that triumphed in 1949, after several decades of war and social unrest. Only after that social revolution did the effect of the agricultural revolution, dietary self-sufficiency, reach all levels of society. Apparently, the tragic effects of food scarcity derived from natural disasters and social disintegration have abated across class lines and geographical regions and probably have been eliminated altogether since the 1960s. Even though problems pertaining to production, transport, supply, and distribution persisted, in the late 1980s there was full confidence in China and elsewhere in the world that the issue of food posed no threat to the future of that nation (Orleans, 1977). That was no small accomplishment, if we keep in mind that one-fifth of the world's population lives in China today.

Corn in China

5

CORN & SLAVERY IN AFRICA

Columbian contact not only marked the beginning of a relationship with Europe. It also marked the beginning of an intense although limited exchange between America and sub-Saharan Africa. From the early sixteenth century until the end of the nineteenth century, nearly ten million African slaves disembarked in America to remain there forever. They brought their languages and cultures, their knowledge and memories, their labor power and almost nothing else. Few other assets and products were associated with this human current, this merchandise, this black ivory, as slaves were called. The presence of African slaves definitively branded and enriched American physiognomy and history. The array of goods flowing from America to Africa was also limited in variety. Nevertheless, the American plants that intentionally or unintentionally arrived in Africa profoundly changed the landscape, diet, and agricultural production of that continent, and, with that, the entire culture and history of Africa.

Despite the magnitude of the human traffic and the momentous nature of its impact, the exchange between Africa and America was restricted, unequal, and distorted. It was not until a third party appeared on the scene—the European colonial powers, promoting and imposing the exchange as a function of their own interests, compensated by the enormous profits derived from that trade—that the relation between Africa and America acquired its true dimension and logic. This did not remedy the inequality or distortion of the relationship between Africa and America. On the contrary, it exacerbated its terms while it also shed an explanatory light on it. Only in that triangular framework is it possible to analyze the origin, the development, and the nature of the exchange between the two continents.

In the mid-fifteenth century some form of slavery was practiced on every continent in the world, including pre-Columbian America. Despite its widespread nature, slavery was a limited and secondary phenomenon, more often linked to the domestic and ceremonial spheres than to the productive sphere. In no geographic region or branch of production was the slave relation exclusive or predominant. Slavery was an expression of the accumulation of power and wealth, but it did not serve either to generate or reproduce those very privileges. It still had not acquired an unequivocal racial character. There were European slaves in the Islamic world and in Europe itself, as in Cyprus and Sicily, where slaves worked in the cultivation of sugar cane. There were Moorish slaves in Europe and in the Near East. Slavery was practiced throughout the many different nations and states of tropical Africa. That same region also provided slaves to Europe and the Islamic world. Although we do not have access to quantitative information on slavery at that time, there is almost universal agreement on its small-scale nature.

After the mid-fifteenth century, when the search began for an Atlantic route from Europe to the Far East, the scale and purpose of the slave trade underwent a radical change. Portugal, which initiated the European penetration of Africa, also gained the upper hand in the trade of human merchandise. Slaves, together with gold, ivory, malaguette, and other African goods, arrived in European markets through the hands of Portuguese merchants, who established the first of several trading posts, fortified and permanent commercial establishments, on the African coasts in 1445. Probably the majority of the slaves taken captive on the coast of West Africa during the fifteenth century were never destined to arrive in Europe at all. Instead, they were used to settle Africa's Atlantic islands that were occupied by the Portuguese, such as Cabo Verde, São Tomé, and Príncipe. There they worked planting and milling sugar cane. The slave plantation became the dominant if not the exclusive organization of production on those islands. The most conservative figures fixed the number of slaves captured by the Portuguese in the second half of the fifteenth century at 30,000. Other sources, working with reliable numbers, increased this number to 150,000. From that time forward, the nature of slavery underwent a transformation in order to become one of the driving forces in the rise and development of modern capitalism.

Portuguese dealers sold some 140,000 Africans in Europe and on the Atlantic islands between 1500 and 1650. Some European slaves, a handful, arrived on American soil, accompanying the first Spanish expeditions. In the New World, one of the most tragic effects of the Spanish conquest quickly became apparent: the precipitous decline and at times the total extinction of the native population. War and the forced labor associated with its aftermath decimated Native Americans. Epidemics especially took a toll. New diseases found fertile ground in the New World population's lack of acquired immunities and in the malnutrition and hunger derived from the dislocation and the rending of the native systems of production and distribution. The lack of labor posed a serious risk that the incredibly vast American territory, whose real dimensions still were unknown at the beginning of the sixteenth century, would become nothing more than a barren wasteland. Without workers there was no meaning, no incentive, and no hope for the permanent colonization of the New World. The substitution of slave labor for indigenous labor emerged not only as a logical strategic possibility but also as a splendid business opportunity. Large-scale slavery served reasons of state that strove for the expansion and consolidation of an empire on which the sun never set. This was in addition to private interests that promoted the massive importation of slaves for reasons of profit. Even religious and humanitarian causes supported demands for black slavery. They were concerned over the sudden extinction of the Native American population, and Friar Bartolomé de las Casas, the passionate protector of the Native Americans, was their collective spokesman.

In 1518, the Spanish crown granted permission for the direct introduction of 4,000 slaves "taking them from the islands of Guinea and from other customary places, and without taking them to be registered at the Board of Trade in Seville" (Aguirre Beltrán, 1972: 17). The permit was sold first to private merchantmen and then to slaving captains, presumably Portuguese. So began the direct traffic between Africa and America that, up until 1870, brought 9.3 million African slaves to the New World. These figures are taken from Philip D. Curtin's *The Atlantic Slave Trade: A Census* (1969), a work based on careful and extensive research. Curtin cautioned that quantitative precision was impossible and that, therefore, his figures only should be used as an indicator of the magnitude of the trade. The figures themselves could vary, plus or minus, by as much as 20 percent. Later research confirmed the accuracy of such a tremendous volume of slaves, although many authors estimated that the real figures were closer to the upper limit estimated by Curtin of around eleven million (Davidson, 1980: 95–101). There never will be absolute certainty with respect to those figures, figures that illustrate the scale of the slave trade but easily could overlook as many as one or two million human beings in the process.

The pace of that gigantic migration steadily increased up until the beginning of the nineteenth century: 125,000 in the sixteenth century, 1,280,000 in the seventeenth century, 6,265,000 in the eighteenth century. Despite the prohibition of the traffic by the English after 1807, followed by the French and other nations, 1,628,000 African slaves were brought to America in the nineteenth century. Slavery was finally abolished all over the American continent in the second half of the nineteenth century. Only then was the human trade gradually brought to a close. Eighty percent of the slaves who arrived in the New World did so between 1701 and 1850, the same period in which the

Enlightenment and the great European bourgeoisie revolutions unfolded, the Industrial Revolution conspicuous among these.

The cultivation of sugar cane and the production of sugar were the activities that were to monopolize most of the slaves who survived the Atlantic crossing. Sugar cane (*Saccharum officinarum* Linnaeus) was not an American plant, but rather was introduced after Columbian contact. Beginning in the seventeenth century, tropical America became the chief productive region in the world for this plant, a distinction it continues to hold. The consumption of sugar in Europe went from being an almost extravagant luxury in the fifteenth century to being one of Europe's basic foods beginning in the eighteenth century. Sugar was one of the most important sources of calories for workers in their struggle to keep up with the new, faster work pace demanded by the Industrial Revolution (Mintz, 1985).

At least two-thirds of the African slaves imported during the long period of the transatlantic traffic were linked to the production of sugar. Their distribution in America reflected that association between sugar and slavery. Brazil is the world's leading producer of sugar, a distinction that began in the seventeenth century and continues today. Brazil acquired 38 percent of imported slaves, more than 3.5 million of them. The British Antilles and French possessions in the Caribbean and Spanish America shared second place, each with 17 percent of the total. Haiti, with nearly 900,000 directly imported slaves, Jamaica with 750,000, Cuba with 700,000—more than three-fourths of those acquired after England prohibited the traffic—and Martinique and Guadeloupe with nearly 700,000 figured prominently among the most important destinations in the geographical logic of sugar and slaves. The United States received 6 percent of imported slaves, nearly 600,000—with fewer slaves exclusively associated with sugar economies—and nearly as many again arrived in the Caribbean sugar colonies belonging to Holland, Denmark, and Sweden.

Those parties directing the transatlantic traffic also changed over time. The slave trade was an attractive source of profit, energetically disputed and perennially subject to piracy and contraband. The Portuguese began the large-scale traffic. They were able to maintain their position of supremacy in the slave trade during the fifteenth and sixteenth centuries. Even though other powers eventually supplanted the Portuguese in this role, the Portuguese remained important dealers until the last days of the trade. The Dutch displaced the Portuguese as the principal dealers in the seventeenth century. In the course of that change, Portuguese slaving captains, independent and mid-sized entrepreneurs, had to compete with large enterprises, true monopolies, that enjoyed state backing such as the Dutch West Indies Company. Large French and English slaving enterprises that began their operations in the mid-1600s dominated the trade before the end of the century. In the eighteenth century, the English once again resorted to independent entrepreneurs, who traded more than two million slaves over the course of that century, three times the volume of slaves handled by their closest competitors, the French and the Portuguese. After the English prohibition, the bulk of the transatlantic trade once again fell to independent entrepreneurs, to smugglers, of every nationality. Other lesser powers, like the Swedes and Danes, appeared on the scene along with the more traditional slaving powers. The Spanish, who in a direct way were lesser participants in the trade by virtue of the lack of trading posts on the African coasts, were an important factor in the configuration of the transatlantic traffic through the sixteenth-century permits and, beginning in the seventeenth century, the asientos, or state concessions for the supply of enslaved labor power. There were no guilty parties, and certainly no innocent ones. There were victims all the same.

As the volume of slaves and their points of destination multiplied, methods for acquiring African slaves changed accordingly. In early slaving expeditions, anyone unfortunate enough to be happened upon was summarily kidnapped. Europeans soon improved upon these methods by entering into alliances with African kings and chiefs in order to undertake joint military expeditions for the capture of slaves. In the course of such operations, it was not uncommon for European dealers to become pawns of local governments or of their opposition. Such occurrences never disappeared entirely but tended to be fortuitous. This type of trade was complemented by the more or less peaceful trade in slaves. The most important mechanism in this peaceful trade was the bounty paid by African intermediaries who were responsible for the capture and delivery of the human merchandise.

African intermediaries were by necessity and by definition political powers invested with military force. At times this meant traditional governments that had assumed new functions. New governments also arose whose only legitimacy derived exclusively from their association with European slave dealers. The intermediaries had the task of distributing European trade goods received as payment, among which firearms figured conspicuously. They also controlled the supply of food to trading posts and the provisioning of slaving vessels for the ocean crossing. They were soldiers, merchants, and entrepreneurs, the African counterpart of the slaving captains who were rewarded with titles of nobility. The Europeans arrived at the points of concentration with their merchandise, proceeded to haggle, at times paid landing rights and other duties to the African chiefs, hastily loaded their cargo in order to avoid contracting yellow fever, and departed for the New World.

The layover on the African coast was only one leg of the trade route revolving around slaving traffic known as the Great Circuit. The journey originated in some European port, such as Liverpool, the most important eighteenth century port in the Great Circuit. There a cargo of metal tools, firearms and powder, textiles, rum, and tobacco was taken on in order to exchange it for African slaves. Firearms played a predominant role and a strategic position in this commercial circuit. It was estimated that in the second half of the eighteenth century West Africa imported over a quarter of a million weapons annually (Wolf, 1982: 210). The going exchange rate at that time was still commonly calculated at one slave for one firearm (Davidson, 1980: 239–46), although figures suggested that firearms were becoming cheaper. Weapons became one of the mechanisms for the spread and reproduction of the slave trade.

European vessels arrived at African trading posts or points of exchange in order to acquire slaves and provisions for sustaining that cargo during the voyage. No money exchanged hands in that transaction. Rather, it was a barter system based on a whole host of agreed upon and probably widespread equivalencies. After the ocean crossing, the human cargo was sold and the ship was loaded with a cargo of sugar, rum, and tobacco. The vessel then returned to the point of departure, where these products of slave labor on American plantations were sold.

European dealers expected three differentiated types of return from their investment in the great slave circuit. The first derived from doubling the price of the European trade goods exchanged for slaves. The second derived from doubling the price at which the African slaves were sold in America. The third derived from the sale of American products in European ports, with private merchantmen expecting earnings 50 percent over the price paid for the American products. The expectation of a return on their investment of close to 600 percent was more a matter of creative accounting than a reality, since the costs of the prolonged operation had to be deducted from any earnings. Even so, profits were substantial and attractive, although there was no consensus with respect to their amount.

The sum total of the economic activity generated by the slave trade, including everything from the manufacturers of weapons and textiles to New World plantations, had enormous importance for the development of Europe and the formation and accelerated accumulation of large capital. Eric Williams, in his classic work *Capitalism and Slavery* (1944), maintained that the slave trade and its aggregate effects provided the capital that allowed the Industrial Revolution to begin. Eric Wolf, using additional resources, tempered that interpretation without contradicting it, by sustaining in a convincing way that the Atlantic trade constituted the "principal dynamic element" for English industrial development (Wolf, 1982: 200).

For the duration of the slave trade, a little more than four hundred years, the stigma of slavery came to include a racial component. Black and slave became synonymous. Distinctive skin color, together with religious, ethnic, and linguistic differences among the Africans themselves, served practical ends in controlling slaves in America. Intertribal differences served to divide the slaves as a group, and the black/white distinction guaranteed that slaves would not be confused with the local population. Racial distinctions lasted far beyond their original pragmatic purposes and hardened into racial prejudice, into an exclusionary barrier. As much in the New World as in the Old, recreated racial prejudice persists.

Then there was the other side of the coin: the effects of the slave traffic on the population and development of the peoples of tropical Africa. This aspect was much less documented and less well known. From the African perspective, the same figures took on new meaning. The 9.3 million slaves painstakingly counted by Curtin were those who lived to disembark on American soil. There were no records to allow for calculating how many embarked in Africa and never arrived at their reluctant destination. Curtin himself (1969), working with an eighteenth-century sample, concluded that 16 percent of slaves perished during the Atlantic crossing and estimated that that proportion could be considered representative of the traffic as a whole. There were documents and dramatic narratives of transatlantic voyages that suggested that such a figure could be conservative. The resounding silence on the thousands of voyages that took place without incident, in which captains simply aimed to avoid the deterioration of the merchandise beyond normal expectations, favored that figure of 16 percent as the best available estimate. This implied that 1.5 million slaves died in the ocean crossing, raising the total number of Africans leaving for America to 10.8 million. Although the transatlantic traffic was the most important, it was not exclusive. African slaves were shipped to the Islamic world and points east over the same period. This traffic, much more cruel and costly in terms of human life according to those who have written about it, increased exponentially after the prohibition of the transatlantic traffic in the nineteenth century and lasted into the twentieth century. No evidence existed with which to quantify that secondary traffic. The most that can be said is that it was not insignificant. Colin McEvedy and Richard Jones (1978) put the number of African slaves exported in the course of that secondary traffic between 1500 and 1880 at 14.9 million.

Those figures, which referred only to exportation, failed to include the human losses within Africa, that is, loss of life associated with the slave trade in the course of capture, transport from the interior to the coast, and the concentration at the point of embarkation, waiting for export dealers to arrive. J. D. Fage (1969: 94) estimated that less than a third of the slaves procured by African intermediaries were acquired through conventional channels, through those mechanisms traditionally recognized as grounds for enslavement: the institution of pawnship, the liquidation of debts, the commission of crimes, or the expulsion from the lineage. Others were kidnapped and, being poor and lacking patrons or resources to buy their liberty, subsequently enslaved. At least half were enslaved as prisoners of war. The spread of war was the most important mechanism for the stockpiling of exportable human merchandise. War caused an enormous loss of human life in Africa, especially after the massive dissemination of firearms. There were no data that permitted the quantification of human losses caused by the slave traffic on African soil. There was a certain consensus that these were large and significant, although they remain largely unknown.

Not every region in Africa nor all African peoples suffered the consequences of the slave traffic in the same way or with the same intensity. Almost no one was left untouched by the direct or indirect effects of the slave trade. A little more than half of the slaves exported to America came from West Africa, somewhere between Cabo Verde and the delta of the Niger River. The majority of the rest, denoted as "Congolese" and "Angolas" in slave ledgers, came from Central Africa, to the south of the mouth of the Congo River, extending from the Atlantic to the Indian Ocean. Lesser quantities were enslaved in the Sudan and East Africa. But the complex networks of military or commercial exchange penetrated far into the interior of the continent, capturing slaves or provoking retreat and migrations in order to elude capture. Captive slaves had to travel hundred of miles before they reached their point of embarkation. The traffic ended up directly affecting or indirectly acting as a threat or restriction against the entire population of Africa and its relations of exchange.

Debate has raged for many years on the demographic effects of the slave trade and the associated genocide. There is no clear consensus, although the most widely held opinion seems to be that, in the long run, the demographic consequences of the trade were marginal (Fage, 1969). These effects, causing population growth to slow to insignificant rates, were compensated for by the

Corn & Slavery

introduction of American plants that appreciably increased the availability of food over the same period in which the traffic developed (McEvedy and Jones, 1978). Nevertheless, it is inappropriate to apply that aggregate and far-flung assessment to specific peoples or to historically specific periods in which the effects of the trade can be definitive for the alteration of demographic development. Neither has it been possible to evaluate yet another element, of a qualitative nature. The slave dealers had a clear picture of the most desirable, highest quality merchandise: the *peça de Indias*, a young able-bodied man in good health, at least seven palms in height, about five feet seven inches tall. The women, children, and older slaves were worth less that a *peça* (Wolf, 1982: 222). This model, implying preferential selection in capture, had to have had consequences in the availability of labor power and its distribution in African households, although we cannot establish all its cultural and social ramifications.

There is consensus on the severity of the political and economic effects of the trade on African peoples, although not all of these have been explored in depth. The trade was established on the basis of preexisting networks of exchange. These, including the exchange of slaves, were much more complex and diversified, based on the explicit availability of labor and the territorially based production of goods. The exchange became specialized and simplified on the basis of the slave trade. It also distorted the entire economic pattern when the exchange was concentrated in only one exportable commodity: this selfsame labor power, which threatened the present and the very future of African productive structures. The trade radically transformed the African economy, African social systems, and African political patterns. Entire African states disappeared and new states arose under the protection of the traffic. Armed confrontation replaced other types of relations. Power relied on external resources and on the role of intermediaries. Nineteenth-century explorers, the vanguard of colonial occupation, and twentieth-century anthropologists discovered what they thought was a primogenitive, primitive Africa. Instead, Africa was in great measure the product of the slaving relation with Europe over the preceding several centuries (Wolf, 1982: 195–231; Hopkins, 1973; Rodney, 1982). Africa was a reflection of its own image, however reluctant non-Africans were to see it or accept it as such.

This brief outline of the development of the slave trade can leave the impression that slavery's expansion was a triumphal process, overwhelming, never slowed by resistance. That impression is false. Although we do not have a clear picture of the resistance to the expansion of slavery available to us, the data clearly suggest that such resistance was constant, with very diverse strengths and with varied manifestations, as much in Africa as in America. The expressions of resistance, characterizing in part the relations between owners and slaves in the New World, were always present in the trade. On the African side this meant the refusal of rulers to participate in the trade, massive withdrawals of entire populations in order to avoid capture, armed resistance, flight and ransoms, rebellions and mutinies in the points of concentration and on slave ships, among others things. Resistance in America took the form of slave revolts, the establishment of fugitive slave communities, and even cimarron republics. In some cases slaves purchased their freedom or were manumitted. Slaves engaged in ongoing demonstrations of passive resistance, from disobedience or theft to feigned stupidity. Resistance was one of the constituent parts of the slave trade. On more than one occasion, such resistance was successful. In many more cases, it was brutally put down and repressed. The slave traffic established its predominance, its hegemony, over Africa, just as the slave plantation established its own predominance and hegemony over vast tracts of the New World, with severe contradictions and despite the ever-present and ubiquitous resistance of its victims. Justice does not always triumph in history or does so too late.

Corn was the principal staple food, the dietary mainstay, for the vast human mobilization brought about by the slave trade. The date that corn was first cultivated and used on the coast of West Africa and its Atlantic islands cannot be determined with absolute certainty. It appears to have taken place sometime within the first forty years after Columbian contact. Such uncertainty derived from the fact that the Portuguese, probably the ones to introduce the plant to Africa in the first place, did not refer to corn at this time by a distinctive name. They called corn *milho*, the same name used to refer to two other cereals that formed part of the former agricultural patrimony of the Mediterranean world: *Panicum miliaceum* Linnaeus and *Setaria italica* Beauvois. To make matters even more confusing, the name *milho* also was used to refer to yet two more cereals the Portuguese discovered in fifteenth-century Africa: *Andropogon sorghum* Brotero and the species *Pennisetum*. The adjectives that modified the word *milho*, which later would permit a clear distinction between those plants, still was not well established, and some chroniclers confused and interchanged them. Authors who, like Jeffreys (1971), looked for the origin of corn outside America or tried to prove that it was introduced to the Old World before Columbian contact, supported their arguments with this Babel of names, despite the fact that the archaeological, historical, and botanical evidence did not support their elaborate arguments.

Corn & Slavery

One of the first unequivocal references to corn cultivation in Africa appeared in a work by Giovanni Ramusio, who published his collection *Dei navigatione e viaggi* in 1550 and 1554. He included a narrative by an anonymous pilot who described the sowing of "*milho zaburro* that in the West Indies is called corn" on the island of Santiago de Cabo Verde. The anonymous pilot's description had to have been made sometime between 1535 and 1550. The citation, well founded, has been the object of some debate, although there are not sufficient grounds to discredit it (Messedaglia, 1927: chaps. 9, 11; Weatherwax, 1954: 35–37; Godinho, 1984). There was a good possibility that two prior references to *milho de mazaroca* on the islands of Cabo Verde in 1528 and 1529 pertained to corn. On the other hand, it seemed doubtful that an item about the cultivation of *milho* on the island of São Tomé in 1506 referred to the American cereal (Godinho, 1984). After 1550, references to corn on the coast of West Africa were relatively common, although these were not frequent enough to actually trace the process of the migration and spread of corn in any precise way.

Various sources attributed corn's introduction on the coast of West Africa to the Portuguese. The Dutch chronicler P. de Maares wrote in 1605 that the Portuguese introduced corn from America on the island of São Tomé and then proceeded to distribute it among the savages. Once they became aware of its abundant yields, the natives adopted corn as a subsistence crop to mix with millet, their traditional staple. The linguistic trail, decidedly elusive, confirms such a scenario. Corn is still known as European grain or, more precisely, as Portuguese grain in several different languages spoken along the African coast. Duarte Lopes noted in 1591 that corn was called *mazza maputo* in the kingdom of the Congo—*maputo* was the name given to the Portuguese—and went on to add that corn was the most vile of grains, scarcely fit for swine. In 1600, corn was called *masinporto* in northern Angola, yet another clear reference to the Portuguese (Miracle, 1966: 93).

Portuguese colonists probably deliberately introduced corn because Old World cereals did not prosper on the Atlantic islands they inhabited, so near the equator, while the yield of African grains was very low. Portuguese settlements and trading posts had to be provisioned with bread flour by sea from the Iberian peninsula for several years. This was fine for Portuguese residents, but too expensive to feed the growing slave population. The 1529 reference to *milho de mazaroca*, very probably corn, mentioned that the grain served as a staple food for slaves. Corn's early introduction into Africa led Vitorino Godinho to speculate that the plant was not brought from Brazil, as generally had been assumed, but directly from the Spanish Antilles (1984). It conceivably could have come from Portugal, where it already was known. There is sufficient literature and precedent to fuel continued speculation about this issue.

There was other evidence pertaining to the early connection between corn and slaves in Africa. Since 1506 slaves from Benin and the *mani kongos*—named after their monarch—were concentrated on the island of Sâo Tomé. In 1533, the vessel Santa María da Luz took on a cargo of 240 slaves and twenty-two baskets of *milho das antillas*—undoubtedly corn—on that island for the Middle Passage. Only 167 of the luckless slaves ever arrived in the New World. At the same place in the same year, the ship San Miguel took on a cargo of 210 slaves and corn for the same voyage (Godinho, 1984). In the sixteenth century, slaves on Portuguese vessels received two meals a day, one of corn and the other of beans—also probably of American origin—salted and cooked in palm oil. They also received only just enough clothing to cover themselves in the interest of modesty. At that time, the Dutch considered such Portuguese treatment of slaves to be an extravagance (ibid.).

By the seventeenth century, corn already was widely and firmly established on the Atlantic coasts of Africa and probably in vast areas of the interior as well. Corn was considered to be a widespread and common staple for slaves, as much on land as during the ocean crossing. In 1682, Jean Barbot observed that between February and August of that year, corn fluctuated in price from a crown to twenty shillings and assumed that the increase owed to the great number of slave ships arriving on the coast. The same author added that farmers saw large profits from corn, selling it to European forts, to slave ships, and to other nations (Miracle, 1966: 91). Corn, besides serving for direct consumption by its African producers, had become a regional mercantile good. Africans met the needs of European slaving consumption with corn and the demands of their own domestic consumption as well.

The commercial corn crop constituted one of the sources of the power and wealth of African middlemen linked to the slave trade. The most famous of these entrepreneurs—the English called them big men—was Johnny Kabes of Komenda, who became the principal intermediary between the Asante and the English in the early eighteenth century. Among his many enterprises, Kabes provisioned the English with raw materials and labor to construct their forts, owned flotillas of canoes for hire, and controlled salt pans and corn plantations that furnished food for slaving vessels embarking on the Middle Passage (Wolf, 1982: 209). This type of middleman and the political entities created through the linkage with slaving, such as the Asante, were efficient instruments in the diffusion of the cultivation of the new American plant. The

Corn & Slavery

peoples living to the north of the catchment of the Congo River attributed corn's introduction to the Asante, who invaded them in the early nineteenth century (Miracle, 1966: 95).

Readily available data confirm that corn was the principal food used in the slave trade. Each slave ship needed a minimum of more than thirteen tons of corn for the transatlantic leg of the voyage, supposing a normal duration of forty-five days and an average cargo of 250 slaves with a daily ration of just over two pounds of corn per person. That figure was extrapolated in a very conservative way from data in the work of Godinho (1984). More than 40,000 transatlantic voyages were necessary in order to transport the enormous mass of slaves to the New World. In the eighteenth century, around 300 ships departed every year, almost one a day. At that time, the slave ships also were known as *tumbeiros*, or tombs. This was a dual allusion to slaves holed up in small niches, like in the ossuaries in mausoleums, and to the human cargo's high mortality rate. The ship's crew was minimal, no more than a dozen or so, and it necessarily imposed the harshest of disciplinary regimens in order to avoid disobedience and mutinies. Abolitionists described the conditions under which slaves suffered during ocean transport in some of their most persuasive arguments against slavery.

There is no basis by which to calculate corn consumption during the mobilization and concentration of slaves on the mainland, although that figure probably was much higher than the amount of corn required for their ocean transport. Miracle (1966: 91–92), using preliminary and rough calculations, asserted that just over eleven thousand tons of corn would satisfy the demand of the slave trade in any given year. That figure easily could be higher. If we accept that figure at face value, it probably meant that the seventeenth- and eighteenth-century market for corn generated by the African slave trade was the largest in history for that grain outside of America. Considering that both a regional and local market existed besides that market linked to the slave trade, corn in some areas of West Africa was, after slaves, the leading commodity and the principal medium of exchange.

Corn was not the only American plant growing in Africa during the time of the slave trade; corn and beans, however, were probably the first. Cassava and manioc arrived several years later and also became important staple foods. Other American plants were grown on African plantations during the colonial period of the nineteenth and twentieth centuries, especially rubber, cacao, sisal, and tobacco. Corn's role in both the slave trade and as a staple gave it a special status and could be explained on several orders. In the first place, corn's high yield set it apart. Corn yields were demonstrably higher than those of traditional African cereals such as sorghum and millet. The difference could not be documented, but it had to have been considerable, since virtually all the chroniclers mentioned or emphasized it. Corn's short growing cycle, shorter than that for sorghum and millet, also constituted an important advantage, especially in those areas with brief rainy seasons.

Just prior to this time, the cultivation of rice had been introduced on the coast of West Africa. Corn had certain advantages with respect to rice in terms of yields and its short growing cycle but, most importantly, corn did not require the technological transformations that rice demanded. We do not know if rice cultivation was introduced in flooded patties or in unirrigated plots, but in any case growing rice required new agricultural methods specific to that plant alone. Rice cultivation spread in a limited and selective way. Even so, it did come to be a common dish for the handful of Europeans living in Africa and of settlers in some areas of the Atlantic coast.

Corn, on the other hand, adapted rapidly to slash and burn agriculture and to mixed planting, which was the most widespread and common method of cultivation in tropical Africa. The corn variety that was introduced in Africa probably came from the Antilles. Systems of cultivation were similar in these two places, which perhaps influenced the Portuguese decision in selecting corn in the first place. Corn adapted naturally to a preexisting tropical agricultural tradition using multi-cropping. Corn responded to standing African agricultural know-how with high yields. The spread of metal implements was associated with the slave trade. These greatly facilitated some of the most onerous tasks of itinerant agriculture and corn probably reaped the benefit. Iron implements and corn were one of the most efficient combinations responsible for increasing the productivity of agricultural labor under tropical growing conditions. In this system, a plot was cleared using fire, cultivated for a reasonable period, abandoned for the time necessary to restore vegetation, and new spaces came under cultivation. There was a symbiosis between the agricultural system created by African cultivators and the corn plant created by American agriculturists.

Corn had no clear advantage in terms of yields over cassava or manioc, but in terms of the slave trade it possessed other favorable traits. The slave trade created the need to store food in order to feed large numbers of people over an extended period of time. This need, so obvious, had no easy solution in the humid climate of the coast of West Africa or during the ocean crossing. On an ocean voyage corn enjoyed a certain advantage over cassava. Corn's kernels were compact and dense. Cassava was a tuber that could be stored in a scattered way on land but that could not be stored for long under other

Corn & Slavery

circumstances. Transport, another of the ineluctable requirements for the mobilization of large numbers of people, was related to preservation and storage. Corn had the advantage of being compact, providing a good proportion of nutrients in a small lightweight bundle. Corn became the food of the porters, the most important means of transport in tropical Africa, and remained so until after the end of the trade. Another of its advantages was corn's high vitamin content that prevented scurvy, the most common and deadly diseases of long ocean crossings.

Mercantile relations functioned to articulate corn and the slave trade. Except on those islands occupied by the Portuguese, and it would seem even there on a small scale, Europeans did not concern themselves with the production of the food necessary for the trade. Rather, they acquired it directly from African producers. We do not know exactly which type of relations of production or exchange African provisioners put in place for the supply of staples. Plantations established for the production of foodstuffs, like those of Johnny Kabes, were one response but not a widespread or common one. The sale of surpluses by family units of production could not be ruled out since the availability of surplus increased with the adoption of the new crop. In whatever form, the mercantile exchange of food on a scale heretofore unknown had repercussions that would be apparent above and beyond the duration of the slave trade. These transformations, together with the exportation of population, branded the development of the economies and societies of tropical Africa, the Dark Continent. Although African exploration had been minimal up until that time, the continent already reflected the impact of imperialism and the mercantile expansion of the European powers.

It is not possible to contend that without corn the slave trade never would have taken place. But neither can the reverse be sustained: that the trade would have found an easy substitute for corn. A sure thing never relies on only one means, but on a combination of incentives and possibilities. Slaving in Africa preceded corn's introduction, but its expansion and exponential growth only can be explained by the incentive of American demand and by the possibility that corn offered to supply and expand it. The slave trade was not destiny or fate, but a series of opportunities and limitations. So, too, those opposed to slavery did not simply appear, but were social groups with the emerging power and will to confront that circumstance. The slave trade was not an aberration, but neither was it the result of a general law of historical development. Rather, it was history: something that happened, but that just as easily could not have taken place at all.

6

CORN & COLONIALISM

In the nineteenth century European explorers began to compile information on Africa. They found corn everywhere in tropical Africa. Only a few scattered areas remained where people still had not adopted corn. In the heart of Africa other peoples readily used the American cereal as their main food, their common everyday staple. This especially was the case in the vast savannas, that region forming a wide horseshoe around the equatorial heart of the tropical jungle in the western part of the continent. More commonly, corn was one among several cereals that together made up the bulk of the diet of those peoples who depended primarily on grain for food. Those peoples lived in an area that more or less coincided with the horseshoe formed by the savannas. Here, on the fringe of the tropical jungle, corn consumption was seasonal, at times limited to tender green ears of early corn.

There was a strong association between consuming corn and raising it. Corn was a leading commodity, but at the same time there was no record of any people who ate corn on a regular basis without cultivating it as well. Porters and caravans, responsible for mobilizing virtually all commercial traffic in the subcontinent, depended on corn as the mainstay of their diet. Corn also remained as the most important food for slaves, who continued to be exported despite the British naval blockade. Corn began to taken on religious functions. It was revered, just like other crucial aspects of life. Corn played a central role in some ceremonies and became part of oral history, of myths, through which African history and knowledge were transmitted from generation to generation. Corn was an integral part of the continent's botanical resources and of the repertoire of knowledge that allowed Africa to persevere and even develop.

We know that the Portuguese were the first to introduce corn on the coast of West Africa. It is unlikely that corn's far-flung geographical distribution on that continent could have been the result of dissemination from one lone point of entry. For this reason, while there is general agreement that corn probably was introduced initially to the continent on the coast of West Africa, others have suggested that there was also another point of entry to the north.

In this scenario, corn came in from somewhere along the Mediterranean rim of Africa and was transported overland by way of centuries-old caravan routes that converged on Tripoli (Portères, 1955; Miracle, 1966). There is plausible evidence to support such a hypothesis. The possibility that the two routes of entry operated simultaneously cannot be ruled out. The process of dispersing corn throughout tropical Africa is complex and still cannot be traced with precision or explained in great detail. In that process, the association between corn and the slave trade tainted corn's mercantile nature, a trait that never has disappeared entirely. The association did fade, however, as corn became increasingly important as a subsistence crop, as a crop grown by peasants in those African civilizations organized around agriculture.

The commercial market for corn and some of its forms of production, such as plantations run by African middlemen, became less important as the slave trade ebbed. The prohibition and persecution of the slave trade by England after 1807, and shortly thereafter by France and the United States, did not completely suspend slave trafficking, but it did steadily and palpably diminish it. The average annual exportation of slaves fell by one-fifth after the prohibition of the ocean-based trade. Only the abolition of slavery after 1865 in those American nations that had continued to permit the practice—the United States and, especially, Brazil and Cuba—put an end to the transatlantic slave trade once and for all. The slave trade with the Islamic world and elsewhere continued. The League of Nations sanctioned the Republic of Liberia for participating in the sale of slaves to the Spanish possessions on the island of Fernando Póo as late as 1931. Spain, of course, was not included as part of this sanction.

The volume of the slave trade fell gradually, beginning in the 1820s and 1830s. Eventually it dwindled until it was insufficient to satisfy either the African need for European imports or the ambitions of merchants who depended on the African trade. Other products began to become part of that commerce instead. Especially important was African palm oil. Palm oil was used primarily in the manufacture of soap, but also as an industrial lubricant. The amount of palm oil exported from West Africa in 1810 was barely a thousand tons a year. Between 1860 and 1900 trade in this commodity averaged fifty thousand tons a year (Wolf, 1982: 330). The commercial relations of exchange between Africa and Europe were profoundly restructured in order to fill the void left by the slave trade, which had been the chief economic linkage in the past.

The effects of restructuring foreign trade and of reshaping trade networks in tropical Africa were intense and complex. Military elites and the state organizations created around the slave trade were plunged into crisis. Some of them persisted, but in an altered form. Others were destroyed by secondary states that arose in the nineteenth century. The qualifier "secondary" was derived from the fact that the expansion of such states was based on imported military technology that permitted rapid growth, but that did not correspond to a specific technical and economic base. That contradiction made such states vulnerable and fleeting as dominant political entities. In the process of the formation and expansion of such secondary states, large numbers of people were mobilized and resettled in tropical Africa. Two such large mobilizations, originating in the southern part of the continent, were the *difaqane*, in which the Zulu empire spread to the north and the east, and the Great Trek, in which the Boers—of European origin—migrated to the north of the Orange and Vaal Rivers. Both events had even greater significance for corn's expansion. Just like the Boers, the Zulus and other Bantu groups had utilized corn since before the time of their migration (Burtt-Davy, 1914). The expansion of corn also benefited from the military expansion of Egypt into the Sudan. The Egyptians depended on corn as a basic staple. Something similar also probably happened in East Africa, where Omani Arabs set up the secondary empire of Zanzibar. Corn, the main staple in the African slave trade, became the mainstay for armed mobilization and military expansion of secondary empires in nineteenth-century Africa (Bohannan and Curtin, 1971: chap. 16).

African foreign trade now revolved around the export of agricultural raw materials and the huge discrepancy between price and volume. This new structure necessitated a different relationship between African producers and European merchants. Such new relations, more direct and less intense, also favored a new form of exchange in which European all-purpose money took the place of barter (Wolf, 1982: 330–32). In West Africa a new group of middlemen, the Creoles, arose: ex-slaves from the American continent and their Christian, culturally Westernized descendants. The Creoles settled in Liberia, Senegal, Gambia, Ivory Coast, Nigeria, and Dahomey. They established themselves in Sierra Leone, with the blessing of the slaving powers. The Creoles there, under British auspices, came to dominate the region. The Creoles derived their power from the Europeans and, just like the secondary states, they lacked an organic basis by which to prevail. Coincidental with the emergence of these groups, Christian missionary activity took off. Simultaneously, Islam, already with a strong African foothold, expanded with unforeseen vigor in West Africa: several holy wars broke out over the course of the nineteenth century. Forces such as these reordered the articulation of tropical

Corn & Colonialism

Africa with Europe and affected thousands of purveyors of raw materials for consumption abroad.

The number of European colonists was small and geographically restricted in the nineteenth century. The Boers, descendants of Dutch Calvinists, settled in the extreme southern part of the continent. Their numbers swelled as they were joined by European coreligionists from elsewhere, the ancestors of the Afrikaaners of what is today South Africa. Some merchants, especially English, French, and Portuguese, began as colonists settling at different points along the coast. This motley multinational ensemble grew to the extent that it was possible to control and prevent yellow fever and malaria. The small group of European settlers, together with the Creoles and Asian expatriates, had an important and influential impact that complicated the transformative and disruptive forces that were bearing down and reorganizing African life. Eventually, all such forces were defeated and subordinated by the direct intervention of European powers. These divided up the continent into colonies between 1880 and 1914. The only exceptions were Liberia, which, thanks to U.S. backing, had the status of an independent republic since the mid-nineteenth century, and Ethiopia. Among the colonial powers, England was the most important and came to dominate half of Africa after World War I. France prevailed over a fourth of the continent, followed by Belgium, Italy, and Portugal, in order of importance.

European colonial rule, which lasted until the 1960s, was varied, as much by the changes that came about over time as by those derived from the politics of the metropolis. The African colonies were subjected to all types of experiments. The British practiced the clearly discriminatory principle of indirect rule, in which African authorities answered to royal administrators. Such a system carried the seeds of South African apartheid, although that was not its necessary consequence. The French and the Portuguese included colonial possessions in the metropolis as overseas provinces operating under special status, an equally discriminatory practice. The most audacious experiment was that of the Congo Free State, a mercantile corporation in the hands of European shareholding capitalists, backed by the power of the Belgian State. The Free State was founded by the initiative and private purview of King Leopold II of Belgium and operated between 1884 and 1908. The African population there was cut in half over those twenty-four years (Dinham and Hines, 1984: 20). On a lesser scale, the British, French, Portuguese, and Germans all transferred governmental powers to concessionary commercial firms. An aftereffect of the wars in Europe was the transfer of the subject states of one metropolis to another, at times with the blessing of the League of Nations or, after the Second World War, the United Nations. The African colonial experience suffered all the problems of the realignment of capitalist hegemony in a century of tumult.

Corn cultivation and consumption spread and grew sharply under this multitude of colonial conditions. Colonial rule implied the direct presence of metropolitan interests and the representatives of those interests working through institutions, businesses, and personal contacts. Colonial administration made way for the intervention of those powers in productive processes as well. We need look no further than European mining companies in South Africa and in Katanga in the Congo; than extractive enterprises in lumber and resins such as rubber and, later, agricultural undertakings; than the construction of railroads and ports. New urban centers, the seats of colonial administrations, and, especially, colonial armies were all expressions of direct colonial intervention in Africa. All these pursuits, whether extractive, industrial, or bureaucratic, required workers. The tasks associated with these required labor that was highly concentrated, extremely intense, and very fast paced, requiring new forms of organization. Although remuneration was quite low, salaries were paid partially in currency. All these developments were unprecedented in the experience of Africa and Africans.

The creation of an appropriate workforce bore little resemblance to the former provisioning system that had been derived from agrarian societies producing their own staple foods. In short order, the colonial powers mustered masses of salaried workers far removed from their previous work in agriculture. At times it was necessary to resort to coercive methods to accomplish this objective. The multitudes of African workers, at times mobilized over long distances, could not be supplied with food through the existing market networks. The provision of food rations by white landowners became paramount for the recruitment of a native workforce. The partial payment of salaries with food rations lasted well beyond its initial function and became a permanent norm for the most important labor relation during the entire colonial era: migratory temporary labor. That owed, in part, to their seasonal character, as in the agricultural tasks associated with plantation labor, or to their limited or singular nature, as in the construction of railroads. But more than anything else, it owed to the express intent of the colonizers to prevent the formation of a mass of permanent salaried workers actually residing near where they worked. Laws, administrative directives, and discriminatory norms segregated the workforce into two sectors. All positions of authority, technical posts, and administrative and policy-making positions were

Corn & Colonialism

reserved for white workers and other expatriates. Nonspecialized, routinized, and physically demanding jobs were left to native workers. There was a great discrepancy in salaries between the two sectors and a strict line that prohibited movement between them. There was only one basic category for the African sector: that of nonspecialized laborer at the lowest possible wage.

Other types of discriminatory regulations prevented workers from establishing themselves in white areas or from effectively settling down where they worked or somewhere nearby. Under those circumstances, all African workers were temporary, at times for periods extending up to several years. They returned to their places of origin whenever they saved enough money to satisfy whatever pressing family or tribal obligation may have forced them to migrate in the first place. At times they simply were able to escape the coercion that had forced them to contract themselves out to begin with. A racist colonial mentality promoted a stereotype of the typical African worker as one who returned home when, after several years of hard work, he finally was able to purchase . . . a bicycle. Low salaries, food rations, and the abuses associated with company stores stopped short any significant transfer of wages paid out by colonial enterprises anywhere outside their orbit and control.

Rations and corn became virtually synonymous in colonial Africa. The provision of food to African workers always had corn or corn flour as its main component and, at times, its only one. In 1907, the Department of Agriculture of British Kenya determined that native employees preferred corn flour to flour made from sorghum. For this reason, new corn varieties were introduced and acreage in corn was increased. In 1908, this same department added that corn not only was more flavorful than sorghum, but its production was more dependable, less vulnerable, and needed only four to five months to mature, as opposed to the seven months required by sorghum. This would allow for two corn crops a year, as opposed to just one sorghum crop (Miracle, 1966: 137). From the point of view of white landowners, corn was an obvious solution since it was cheap, easy to transport, and could be stored with minimal spoilage. Wheat and rice were more expensive and tubers and plantains were not easy to transport or silo. Millet and sorghum were not readily available and they were more expensive than corn, despite the fact that producers were paid prices very similar to what they received for corn (Miracle, 1966: 133).

The colonial governments also used food rations for other purposes. African troops recruited by the colonial armies depended on corn in their rations. War in Europe automatically translated into war between the African colonial possessions belonging to the opposing metropolises. There were no great battles fought in Africa in the interest of war in Europe, but those European conflicts did occasion a massive recruitment of African troops, and that in turn substantially increased the commercial demand for corn. Some areas under colonial jurisdiction that had been modest corn exporters, such as West Africa and Kenya, withdrew from the export market in order to attend to domestic military demands during World War I. During World War II, high prices in the British colonies stimulated corn cultivation, once again in order to satisfy military demands (Miracle, 1966: chaps. 9, 10). The association between corn and war, long-standing in tropical Africa, persisted there for the entire period of colonial rule, except for that lapse when the Africans had no reason to fight. War, just like colonial administration and so many other things, was just one more imported commodity.

Colonial governments also saw themselves obligated in some cases to distribute corn rations among the native population at large after insects or drought decimated crops. Bad harvests and insect infestations predated colonial times. Only rarely had widespread famine loomed, however. Colonial rule dislocated native productive systems and destroyed the historical responses of Africans in the face of agricultural catastrophes. Colonial administrations were not always able to avert famine, despite limited attempts to distribute food during such times of crisis.

The widespread use of food rations promoted the expansion of corn production and consumption. The introduction of corn among some peoples was attributed to returning migratory workers who had acquired a taste for corn and had grown accustomed to eating it while they had been away. The expansion of corn production in Zambia was so fast paced that in 1955 yields were twenty-two times what they had been in 1920. Corn became Zambia's leading commercial crop. Such an occurrence clearly was linked to European investment in the development of copper mines in Zambia and Zaire. The objective of such initiatives was to extract African natural resources in order to export these raw materials to the metropolis. The native workforce required for such an endeavor created a whole new demand for corn.

Part of the demand for corn was met by a new type of agricultural producer: the white settler. Private European settlers were already in the region for a whole host of reasons and their presence was part and parcel of colonial rule. Sometimes white settlers were there by design of colonial governments, and at other times their presence was a simple fact of life that could not be circumvented. For many of those settlers corn cultivation was their only alternative if they were to remain in the African colonies at all. Perennial crops monopolized by large capitalist plantations required investment, time, econo-

Corn & Colonialism

mies of scale, and linkage with the world market. They were out of the reach of individual colonists who lacked substantial capital. Stock breeding, a livelihood preferred by many white settlers in South Africa, was impossible to pursue in tropical Africa, due to the endemic presence of the tsetse fly and other diseases affecting cattle in the region. Few annual crops could be successfully propagated and none of them promised corn's high yields or corn's broad commercial market. The very qualities that made corn a suitable crop for the purposes of colonization—its resistance to disease, short growth cycle, versatility, low thresholds of investment in capital and labor, high yields— also made corn the most important commodity grown by white farmers in Zaire, Kenya, Angola, Mozambique, Zimbabwe, Zambia, Madagascar, and elsewhere in the new colonial landscape (Miracle, 1966: chaps. 9, 10). White settlers who commercially cultivated corn brought about important technical changes. The introduction of the plow and draft animals led to moderately higher yields and appreciably reduced labor inputs. Despite such innovations, corn production by Europeans living in Africa never lost its extensive character and the attendant low costs and low yields. Increases in production essentially were a consequence of bringing new lands under cultivation. This was the most economically rational option in circumstances in which the land was cheap and abundant, as it effectively was for white settlers.

Colonial governments always protected white agricultural settlers and bestowed them with many other benefits as well. The most important of these was generous expanses of free or virtually free land. White settlers typically received the lands best suited for agriculture in plots of at least 120 acres and in most cases of nearly 250 acres. That land had not been empty or idle. African farmers had been occupying that land precisely because it was the best land. The Africans cultivated the land in itinerant parcels of just over seven acres apiece. Those African farmers simply were stripped of their lands by virtue of decrees issued by colonial governments. At times, as in the Congo and South Africa, discontinuous cultivation of land was sufficient cause for Africans to lose their access to plots and for the land to pass to the colonial government. Colonial administrators suffered a blinding cultural inability to comprehend African agricultural systems on the one hand, but benefited greatly from a remarkable clarity in their own strategic purposes on the other.

Another benefit for the white settlers was cheap native labor for almost all agricultural tasks. Frequently, native labor was recruited by coercive means with government authorization or by its militarized police forces. The low price of labor was not the result of abundant supply. Rather, it was the result of demands by colonists, who created necessary supply through coercive means. In many colonial jurisdictions the price of corn was controlled through government granaries. The government purchased corn at attractive prices and only European settlers had access to that protected market, which was off-limits to African peasants. They were instead forced to sell at low prices, frequently to a settler who proceeded to take it to a government granary (Miracle, 1966: 253, 262–63). A range of technical services and credit was available to the white settler, but not to the African peasant. The white settler who produced corn was protected by government policy and was in an advantageous position with respect to the African peasant by a simple but determinant fact of the colonial regime: national origin and skin color.

Despite the advantages and subsidies extended to European settlers for corn cultivation, the largest part of corn traded in tropical Africa came from African producers. Only in Zambia and Zimbabwe did European settlers supply the bulk of corn for market. In the 1950s, this meant that anywhere from half to two-thirds of commercialized corn was produced by white farmers. In the rest of the region, including areas where corn was highly commercialized, such as the Congo, Kenya, and Mozambique, African peasants dominated the market. African peasants planted and ate much more corn than ever wound up in European grain markets.

The colonial governments directed almost all their efforts and resources toward the development and growth of corn production by white settlers. This was in spite of the overwhelming numbers of African producers and the discriminatory practices forcing African growers to sell their grain at below market value. Colonial officials justified this preferential treatment and the subsidies derived from it by citing the unstable character of production by African peasants. That is, corn production could flood the market one year and leave it woefully undersupplied the next. Fluctuations could be on the order of 100 percent from one year to the other (Miracle, 1966: chaps. 11, 12). Colonial officials attributed such unstable production to technical shortcomings in African agriculture, which in turn made crop yields more vulnerable to meteorological risks, and to racially determined traits, such as the apathetic, laconic nature of African peasants.

Colonial administrators exaggerated the random nature of production. In the long term, the volume of corn marketed by the African peasant was not only constant, but growing, despite any annual fluctuations (Allan, 1971). Colonial administrators simply did not appreciate or were deliberately oblivious to the real causes behind uneven production. Some of the fault lay with the severe restrictions colonial administrators had imposed on agriculture and land use that were incompatible with African tradition. Discriminatory practices against Africans with respect to prices, access to capital, and other modern factors of production and marketing also contributed to the situation. The single most important cause of fluctuating production levels was the resistance of African peasants themselves. They gave first priority to their own subsistence rather than to supplying corn to those who dominated them.

African nations and peoples suffered tremendous territorial expropriation under colonial rule. European plantations for the production of export products were one of the most powerful agents and the greatest beneficiaries of such expropriations. Plantations, developed in the New World around slave labor, arose in Africa in the first half of the twentieth century in order to cultivate cacao, coffee, tea, rubber, palm trees, sugar, tobacco, and henequen, among other things. These plantations appropriated the best lands to serve their own purposes. Plantations also used vast expanses of land, much larger than their needs and potentials might dictate. This was contrary to common sense, but not contrary to the megalomaniac ambition of the corporations that owned them.

In 1911, colonial authorities in Zaire granted over 180,000 acres of land to a subsidiary of W. H. Lever. The company was to use this prime land for groves of African palm trees. The original grant could be increased to up to nearly half a million acres after ten years. After twenty-five years, the land grant could rise to more than 1.8 million acres and ownership could pass to the company outright. Unilever, heir to that and other land grants, was the largest enterprise with African plantations and in the 1980s was the world's largest multinational producer of foodstuffs. In 1926, the government of Liberia gave Firestone a land grant of up to a million acres for the planting of rubber trees. The price was a mere four cents per acre of land actually sown. Firestone planted only eighty thousand acres of trees, just enough to establish the biggest rubber plantation in the world and to obtain the monopoly on that resin in Liberia (Dinham and Hines, 1984: 21). Typically, such land grant plantations fenced almost all the land concerned, not just land directly under production. Fencing not only deprived the African people of magnificent lands, but also disrupted territorial continuity, closing off passes and routes for itinerant territorial use that exploited land intermittently in long-term cycles.

Expropriations for the benefit of white settlers probably did not reach the scale of those carried out for the benefit of large plantations. Even so, in parts of Zambia and Zimbabwe, Kenya, and in all of South Africa, the territory controlled by the small minority of white settlers, no more than 2 percent of the total population, was greater than that controlled by African peasants, who made up around 95 percent of inhabitants (Miracle, 1966: 31). The expropriatary action of colonial governments went far beyond the excessive tracts of land given over to plantations, settlers, mines, lumber companies, and other European businesses. Railroads and roads, just as other public works, served as a pretext for disproportionate expropriations. More expropriations, again excessive, were carried out for the creation of territorial reserves. Africans were disenfranchised and then barred from access to the largest part of their own territory.

Land policies of colonial governments were not only aimed at benefiting white interests, but also at serving another, self-evident purpose: to confine the native population to small, strictly delimited areas. The confinement of the population served several ends: vigilance and control, tax collection and labor recruitment, education and Christianization, and segregation in order to avoid racial contamination. It also served to keep Africans poor and, with that, put them at the ready disposition of whites as workers or as soldiers, as simple instruments of colonial purposes. Many colonial governments followed the South African model. They established a local legislative assembly exclusively for whites, created territorial reserves for the native population, and decreed that black Africans could not live or work on the land outside a specified area. Native reserves in South Africa scarcely made up one-tenth of the total national territory. With minor variations, that model was reproduced all over the subcontinent during the colonial era.

Expropriations and the enclosure movement severely affected African agriculture and stock breeding. In some cases, the loss of territory made pastoral activity virtually impossible. Such was the case with the Kipsigis of Kenya, who primarily were pastoralists and, secondarily, agriculturists. Authorities restricted the Kipsigis to a territorial reserve insufficient for the stable development of transhumant nomadism. Around 1930, British colonial authorities pressured and otherwise induced the Kipsigis to turn from livestock to agriculture. Incentives included the introduction of plows and oxen for use in cultivating corn. One of the consequences of that circumstance was that fencing, which before only had served to protect crops from livestock, came to mark the boundary lines of incipient private ownership of agricultural land (Manners, 1967). In other cases, expropriation and enclosure broke with the natural cycle associated with slash-and-burn agriculture in Africa. Land had to be cultivated more often, with lower yields and higher investment in labor in order to eliminate competitive underbrush. The incidence of insect infestations also probably increased, to the extent that cultivated fields were in closer proximity to each other. In the 1920s and 1930s, locusts wreaked havoc and devastated crops to a degree beyond recent memory. Soil erosion brought about soil degradation in portions of the savannas and the rainforest. Under such circumstances several factors worked toward creating a preference for growing corn: corn's higher yield compared to other agricultural crops; its greater resistance to some insects; and the height of corn, which outpaced that of competitive underbrush after a certain stage of growth and eliminated the need for some weeding. These characteristics favored a transformation of this American cereal into an effective technical option to cope more effectively with the severe repercussions that territorial enclosures had on the African peasantry.

In other cases, colonial powers set up completely separate governments and administrations within the same jurisdiction. The purposes of such a system derived from one culture and the style and norms of such administrations derived from quite another. The interests of the metropolis dictated separation, but the interests of Africans dictated the character and means by which such an arrangement could be carried out.

Colonial government and administration had a cost that had to be paid for by the inhabitants of these colonies, preferably by the natives, whether they received services or simply tolerated them as boon or bane. The recovery of that cost was the first task for colonial administrations. Decisions about many colonial policies or actions were a function of government income. The best way for governments to generate revenues was to levy taxes on foreign colonial trade. This was easy and economical to regulate because it was concentrated in one or at most a handful of ports (Bohannan and Curtin, 1971: chap. 19). Foreign commerce was preferentially catered to. The entire African colonial economy bowed to it. To this end, imported substitutes displaced perfectly good local products. The stereotype of reluctant native consumers came into play here. Missionaries and civil authorities were ostensibly so concerned about the physical and moral well-being of the natives that they made heroic efforts to eradicate the consumption of locally made thick African beers. These beers typically had a low alcoholic content and almost always were made from fermented corn. Meanwhile, government-subsidized merchants who were worried about the financial well-being of the administration dedicated themselves to promoting the consumption of imported beer.

Taxing foreign trade had its limits, however, and provoked a powerful backlash on the part of the import-export merchants in this sector. Almost all these were Europeans or expatriates, and they had a great deal of influence with individual colonial governments. Such revenues only partially covered the cost of governmental bureaucracy, characterized as it was by an inexorable tendency to expand. Other sources had to be tapped from the start in order to make up the shortfall. Colonial administrators considered a tax on the native population to be the most logical and fair alternative for revenue. It was only fitting to tax those who ultimately benefited from the rational and enlightened management of public affairs, even though the miserable creatures might not appreciate it as such. But it was no easy task to levy taxes on such a widely dispersed population, neither properly accounted for nor registered. Africans spoke one or even several other languages, all equally incomprehensible to colonial officials. The natives so physically resembled each other in European eyes that they appeared virtually indistinguishable. Besides which, Africans behaved in a way that was neither economically rational nor productive. For all these reasons, colonial administrators opted for simple tribute: the periodic payment to the government of a fixed amount of money per person. The amount was independent of wealth or occupation, simply determined by the more or less objective circumstance of being a native. The practice was efficient and had many advantages. The chief effect was that the natives had to obtain money in order to pay the tax, and for this they had to sell something, either a product of their labor or their labor power itself. This directly or indirectly benefited the entire colonial system and imparted a little Western rationality to the confusing and incomprehensible day-to-day existence of the natives. The imposition of tribute or taxes on the native population was one of the first measures taken by colonial governments all over Africa.

In order for colonial governments to justify their cost and their very existence, they had to make themselves readily available for repression on behalf of their home governments. This was, after all, their essential function. Their most important efforts were dedicated to seeing that European interests in Africa were consistent with those of the metropolis. The construction of railroads, directly by the governments themselves or by private businesses backed by European capital, was one of the most flagrant examples of the intense nature of the linkage of colonial governments in tropical Africa to the metropolis. The blueprint for rail lines was clear: emanating out from mines, plantations, and other European capitalist business ventures directly to ports for embarkation abroad. The shape and the borders of many independent African nations today reflect that imperial design of transportation networks (Bohannan and Curtin, 1971: chap. 19). Railroads suddenly provided cheap transport that destroyed and reorganized old networks of exchange. The railroad changed both the products that could be mobilized and their destination. Railroads promoted corn production by white settlers to supply government granaries. They did this by establishing a cheap fixed rate for transporting grain produced by these white settlers, independent of its point of origin

Corn & Colonialism

or its ultimate destination. Corn was a strategic product in the overall colonial plan, the sustenance of salaried labor.

Tribute, salaries, and markets steadily increased pressure for the monetarization of African rural life. Through that process a growing part of peasants' productive surplus passed through the hands of white sectors and, from there, on to the metropolis. When that surplus did not materialize or failed for whatever reasons to enter into African economic planning, monetarization stepped in as the force behind the creation and growth of surplus. It was necessary to generate more and more money in order to continue living in conditions that were the same or even worse than before. The need to obtain money, the veil for the transfer of surplus and labor, was seen as a powerful pressure on territorially bounded African agriculture. The introduction of commercial crops may have produced no goods for direct consumption, but it did produce money. It became necessary to devote a portion of agricultural land to the production of money at the cost of the production of food.

On the coast of West Africa, at the initiative of African producers there, cacao and palm became commercial crops before colonialism's heyday. Peanuts and a few plantation crops, such as coffee, were adopted as commercial crops in rural Africa, altering the structure of production. Where the adoption of commercial crops met native resistance, colonial governments used compulsion or questionable inducements. In the early twentieth century, Belgian officials compelled natives in Rwanda and Burundi to plant potatoes, sweet potatoes, and cassava, or else run the risk of punishment. Around this same time, English authorities in Malawi ordered administrators living in the district to coerce African cultivators to plant corn and meet a quota established for every household (Miracle, 1966: 132, 156). Colonial rule saw the practice of obligating peasants to sow commercial crops repeated again and again. Such authoritarian policy apparently had little effect in guaranteeing a ready food supply, but it did have demonstrable and complex effects on African agriculture.

Labor itself had to be mobilized in order to raise a surplus in those places where natural conditions or great distances with respect to markets made it difficult or impossible to adopt commercial crops. The departure of contingents of migrant laborers, overwhelmingly young men, altered the distribution of tasks in domestic agriculture. Women took on a growing proportion of the work in the production of foodstuffs in vast areas of tropical Africa. In other regions, the absence of these migrant workers was keenly felt and existing plots had to be cultivated for longer than normal periods in order to postpone the backbreaking work of clearing new lands. Territorial limits and the pressure to earn money dislocated traditional systems of production all over colonial Africa.

African land tenure systems suffered enormous pressures and did not go unscathed. It is not possible to generalize about landownership over such a large and culturally diverse area, except with sweeping categories that at times obscure the nature of the relation between society and land. In precolonial Africa, with few exceptions, social relations or membership in a group guaranteed productive units' access to land and set territorial boundaries. Individual owners did not occupy the land. Groups did. There were many variations on a theme within that broad system of tenure. Colonial governments understood none of them. Instead, they reduced a host of complex relations to a single legal rubric: communal property. Colonial officials assumed the group to be, collectively, the sole owner. Authorities defined ownership exclusively in the Western sense of the concept, as a delimited territory with fixed borders (Bohannan and Curtin, 1971: chap. 7). Such a conceptual rigidity fixed boundaries for a delimited area and excluded a whole array of other rights associated with tenure. Such a definition nicely complemented the other colonial land policies of expropriation and enclosure and even created an artificial scarcity of land in some areas.

Colonial powers put so much pressure on native Africans to become part of a money economy that group ties were effectively weakened. Obtaining money was the responsibility of individuals, who were responsible to and penalized by the government if they did not comply. That individual pressure had to be relieved in some sort of collective space. Responses to such a contradiction were multiple and complex. In West Africa, land largely retained its communal character. There were some exceptions. Coffee bushes or cacao trees, commercial crops, were legally recognized as private property for individual gain, and as private property when exchanged for money (Stavenhagen, 1975). Among the Kipsigis of Kenya, permanent and exclusive family parcels were created for corn cultivation in an array of sizes (Manners, 1967: 287–99). Differences between individual members of landowning communities emerged and became increasingly evident. Stratification on the basis of monetary wealth differentiated members and altered social relations all over Africa. Outside forces had been applying pressure on individual communities to privatize for many years. Similar pressures began to arise from within particular communities. The problems surrounding landownership in tropical Africa still had not been resolved in the 1980s. Nevertheless, there was no evidence of an inexorable movement away from traditional forms of land tenure toward private property.

Corn & Colonialism

Corn cultivation became an important resource allowing African peasants to confront the enormous pressures imposed by colonial rule in many parts of Africa. Corn was both a commodity and a dietary mainstay. Few agricultural products possess such flexibility, such a dual nature. The peasant who sowed corn had options and he could render decisions that were closed off to him with other crops. Corn allowed peasants to construct a line of defense for their physical survival and provide for their social reproduction. Peasants tried to both guarantee what they needed for their own sustenance and to earn money by planting corn. Whether the harvest was bountiful or whether it failed determined the degree to which each particular purpose would be served: their own survival or the money economy. As colonial restrictions made agriculture increasingly risky, African peasants gave priority to their own subsistence. Colonial administrators interpreted such choices as unacceptable fluctuation. It seemed never to have occurred to them that the problem could be solved by working to lower peasants' costs, by eliminating and ameliorating some of the restrictions suffered by African producers. This amounts to poetic justice at the end of an era: corn, probably introduced to Africa to serve European interests, became one of the secret weapons in peasant resistance to colonial rule.

: 7 :
CORN & DEPENDENCY IN INDEPENDENT AFRICA

The independence of Ghana in 1957 heralded the massive decolonization of Africa that would take place during the 1960s. Decolonization was a complex process, subject to decisions already made by the colonial powers themselves and designed to preserve the interests and spheres of influence of those same powers. Despite such obstacles, the majority of sub-Saharan African states rose up over the course of the 1960s to become independent. The Portuguese colonies lagged behind the rest and did not acquire their independence until the 1970s. By the 1980s, the governments of almost all African states were in the hands of their native populations, Africans themselves. South Africa was the only exception, with its selective white independence, and even it would eventually give way. Modern African states display an array of government and economic models: de facto military rule, one-party systems, more or less impeccable Western style democracies, varied alignments with world powers, development projects designed to lead the country to capitalism, African-style socialism. The key to the search for Africa's own model for development resides in this very diversity. The plurality that colonial rule tried so hard to suppress emerged in all its splendor and all its contradictions, complete with missteps, setbacks, and dreams.

In the face of such diversity, problems with collective purpose appear and persist—problems derived from colonial rule and from dependence, problems that are a direct result of the relation between newly independent states and world powers. These problems are the subject of this chapter.

The new African states inherited a burdensome legacy from their colonial past. This legacy affected all aspects of life and closed off alternatives for autonomous development. One of the most dangerous and onerous legacies was that of dietary dependence. The population of Africa grew, despite forced migration and high mortality derived from the slave trade and from war and other calamities associated with it. Demographic growth had been steady since the sixteenth century, although only at moderate rates until the nineteenth century. Those rates rose in the second half of the nineteenth century and became explosive in the twentieth century. Food produced in Africa itself provided the fuel for population growth on that continent. The introduction of American plants as an integral part of African diets played an important although still little-known part in that process. Corn and cassava figured prominently, and by the twentieth century provided more than half of all calories in African diets. In the years leading up to World War II, Africa was a modest net exporter of cereals. Between 1934 and 1938, Africa exported an annual average of over 730,000 tons of grain. Corn accounted for nearly 700,000 tons of this amount, which represented something around eight pounds per capita a year (Brown, 1971: chaps. 5–7). South Africa was a leading exporter of corn at that time.

The figures on African exports of basic foodstuffs for those years showed self-sufficiency, but not abundance. The average consumption of cereals in Africa was considerably less, by at least a fourth, than that of the other continents or statistical regions and less than half of what it was in Western Europe. Pioneering studies on nutrition in Africa revealed significant deficiencies in terms of both absolute and seasonal amounts. Absolute deficiencies were established by comparison with norms that determined the amounts and components for adequate nutrition, although the validity of these norms was subjected to severe criticism. Seasonal deficiencies were established by comparing diets at different times of the year. The results showed that in the months just prior to the harvest there was widespread famine (Richards, 1939; Gouru, 1959: chap. 7; Miracle, 1966: chap. 8).

Colonial administrators, doctors, and even scientists frequently and harshly criticized African diets and implied that the natives themselves were responsible for any nutritional deficiencies those diets might imply. Cornmeal mush was a ubiquitous and leading component of most meals. This created the impression that African diets typically enjoyed little variation, that they were inadequate, and that they lacked essential dietary elements. It was believed that this, in turn, led to poor overall health, small stature, reduced capacity to work, and learning disabilities. To these judgments, which were presumably objective, other indictments were added: African food was monotonous, it was visually unattractive, it was prepared under filthy conditions, it was insipid or simply repugnant. Such judgments were based on cultural comparisons. It would have been enlightening to ask those who ate African food everyday what they thought of it. Dietary preferences are a very personal matter. There is no evidence that allows us to endorse such derogatory judgments as objective assessments. There has never been any proof of intrinsic shortcomings in the balance and composition of African diets. Such value judgments cannot clearly distinguish between two very different types of problems: dietary building blocks versus the availability of foods to satisfy them.

Studies on nourishment in colonial Africa almost never took into account restrictions imposed on food systems. Instead, they tended to attribute any obvious dietary deficiencies to agriculture and dietary preferences, to culture. There were demonstrable problems: insufficient access to land, inadequate volume of the production of foodstuffs, the low proportion of food that producers could reserve for their own consumption, a lack of resources to preserve and store foodstuffs, obstacles preventing access to complementary foods obtained from pastoral activity or from hunting and gathering, and forced migration of labor. The result was the pronounced and frequent scarcity of food. The origin of issues such as these never was found in African culture. Rather, these issues were the direct result of the destruction of native dietary systems. It was a complex problem that did not allow the simplistic designation of one abstract guilty party. It was not possible to attribute all the nutritional and dietary problems Westerners observed in Africa to colonialism alone. Even less could the blame for the problems of malnutrition be laid on Africans and their native diets, on their culture.

At the end of World War II, just as in other regions the world over, the directional arrow for exchange of basic staples changed and became unfavorable for Africa. An average of more than 360,000 tons of cereals were imported to Africa every year between 1948 and 1952. That number rose to nearly 1.4 million tons between 1957 and 1959. For 1960–61, on the eve of independence, cereal imports to Africa rose to nearly 2.4 million tons, or an average of almost twenty pounds per person. Wheat, more than 3 million tons of it, accounted for almost the entire trade deficit. The purchase of wheat revealed a profound and induced change in the diet of some sectors of the African population. In Africa, with very few exceptions, conditions favorable for the production of wheat did not exist. In contrast, corn exports continued and even rose moderately, reaching just over a million tons in 1960–61 (Brown, 1971: chaps. 5–7). The import and export of foodstuffs appeared more linked to foreign trade alliances and interests than to any dietary needs of the population at large. Africa's change of course from cereal exporter to cereal importer was not a linear result of demographic growth. Rather, it was accelerated by increased consumption by some privileged sectors of the population.

As the years passed, dietary dependence in Africa worsened, characterized by its uncommon severity and extent. Nutritional deficiencies also became more pronounced for large sectors of the population, perhaps even the majority. Since 1960, per capita food production in tropical Africa declined in three out of every four years, with the predictable result that the average per capita production of foodstuffs in 1980 was 20 percent below that same figure for 1960. Nowhere else in the world was this the prevailing tendency. There was a dramatic increase in imports. In the early 1980s imports accounted for between 25 and 30 percent of aggregate production for the entire continent and some $15 billion a year. Despite this, per capita consumption fell in proportions similar to those for production. In some extensive regions, like the Sahel, the drop in consumption was more serious and there were famines on an unprecedented scale. Dietary dependence, with the exception of a handful of countries, was a growing and widespread phenomenon in newly independent Africa. Some estimates suggested that if there were not a significant increase in domestic production, imports would have to double or triple during the 1990s (Dinham and Hines, 1984: chap. 6).

Explanations for the tendency for dietary conditions in tropical Africa to deteriorate often reverted to data on demographic growth. The subcontinent had the fastest rate of population growth in the world in the late 1980s, at an annual rate of nearly 3 percent. Population growth did play an important role in the dietary woes of tropical Africa, but it could not account entirely for such a complex phenomenon. In comparison, Africa as a region had the least amount of acreage under cultivation with respect to the potential area that could be used for agricultural purposes. These rates in Africa were far below those calculated for Europe, Asia, or Latin America (Grigg, 1980: chap. 18). On the other hand, growth in long-term aggregate agricultural production kept pace with population growth and the growth rate for some crops consistently surpassed population growth by wide margins. The cultivation of sugar cane and tea doubled between the late 1960s and the late 1980s. The cultivation of coffee, tobacco, cacao, cotton, and other export products grew at an accelerated but erratic pace, in response to fluctuations in the international market (Dinham and Hines, 1984). The potential for accelerated growth of agricultural production in tropical Africa was there, but seemed linked almost exclusively to commercial export crops. The relation between territory and population, the old Malthusian dilemma, could not explain away dietary dependence in this instance.

The importance of export agriculture for tropical Africa was overwhelming by the late 1980s. The majority of African countries earned more than half of the total value of their exports from the sale of agricultural products to overseas markets. Seventeen of those countries were monocultures and increased earnings by selling just one agricultural product abroad. Nine countries received more than 70 percent of their foreign exchange through the export of only two to five agricultural products. Only a few countries were able to draw on mineral resources, petroleum, or other raw materials that they exported in an unrefined state. Production in the agriexporting sector was in many cases under the direct control of large foreign-owned plantations. Even though there were exceptions, foreign-owned plantations not only prevailed after independence but grew in terms of total acreage and in terms of their power and influence. All export crops were under the commercial control of large multinational corporations. Even though their production might be in the hands of peasants, as in the case of coffee and cacao, these large concerns monopolized access to world markets and set prices and conditions of sale (Dinham and Hines, 1984). Export agriculture controlled by multinational corporations was a force to be reckoned with and domestic African agricultural policy was completely subordinated to its interests. Public investments, public resources, loans, and foreign aid were all disproportionately aimed at supporting and serving the agriexporting sector. Little or nothing was left over to promote the domestic production of foodstuffs. With few exceptions, scarce resources were concentrated in massive modernization projects influenced by the interests of the large multinational corporations. Their influence in the conception, design, and financing of such projects promoted a clash with traditional agriculture and an increasing degree of dependence. The production of foodstuffs in almost all African states was relegated to the politics of development.

The United States encouraged African dependence on imported food during the 1960s. The United States supplied cereals and other basic foodstuffs at low prices on terms that were financially attractive but politically costly. It was or at least it seemed easier and more rational to import food and to develop the agriexporting sector in order to pay for it. Grain imported by countries in tropical Africa doubled during the 1960s, but their cost rose by less than 50 percent (Dinham and Hines, 1984: chap. 6). Part of that cost could be paid in local currency under Title One of the Agricultural Trade Development and Assistance Act of 1954, or U.S. Public Law 480. Offers of foreign aid, almost always associated with wheat, were not always consistent with the needs of traditional diets. Massive wheat imports profoundly changed the dietary habits of urban populations in Africa, who were after all the targets for the consumption of those imported foodstuffs. The milling and baking industries arose as a result of wheat imports and were often in the hands of many of these same multinational corporations. These new sectors came to play strategic political roles by virtue of their direct connection to urban consumers.

Corn & Dependency

Little or no imported foods ever made it to rural areas, but the effect was felt in the depression of domestic prices that, in turn, were tied to the prices of imports.

Violence during the 1970s disturbed many of these conditions but, in general, African countries became increasingly dependent. The price of imported foodstuffs rose rapidly after 1972, when the United States reopened grain sales to the Soviet Union. On the other hand, the prices of African exports dropped and experienced extreme fluctuations. The agriexporting sector needed to be propped up in the face of deterioration and uncertainty. Prices of imported foods for the urban population were subsidized in order to avoid excessive price hikes and avert any risk of an adverse political reaction in the seats of power. Subsidizing imports kept domestic food prices low. Any attempts to increase domestic food production faced enormous obstacles. Impediments ranged from design and development issues, to a scarcity of public resources to finance the implementation of new ideas, to ways to link domestic production to urban markets already accustomed to imported food. It was the worst of all possible worlds. Dietary dependence in tropical Africa showed no sign of reversing itself, but neither did it seem possible to increase domestic production in the face of scarce foreign exchange. It was a vicious circle.

In the midst of this crisis, corn cultivation appeared on the scene as a glimmer of hope, but also as a risk. Corn production in Africa over the period from the late 1960s to the late 1980s had not followed the same decreasing tendency as other basic staples. Neither had corn production increased sufficiently to significantly affect Africa's dependence on imported food. According to available statistical information and excluding South Africa, corn production in the rest of tropical Africa rose from a yearly average of 9.3 to 16.5 million tons between 1960 and 1980, a rate slightly higher than that for population growth at large. Here I make some necessary asides. First, the quality of agricultural statistics for the African countries is very poor, especially those referring to the production of basic foodstuffs. The figures I have used, taken from sources published by the Food and Agriculture Organization of the United Nations, are not exact. Nevertheless, they serve to illustrate matters of scale and relevant trends. Second, it is important to note the exclusion of South Africa, the leading producer of corn in Africa at that time, accounting for about a third of total production. Corn production was carried on in that country mainly by white settlers and had a mercantile character. South African corn exports prior to the dismantling of apartheid had no bearing on imports elsewhere in Africa but, rather, were integrated into the world market. South African corn production was not part of African food systems at large. It did have enormous importance for the diet of the African population of that country, which it tapped through commercial channels of distribution.

Corn production grew so much in tropical Africa during the 1970s and 1980s that it became the leading food crop for the entire subcontinent. By the late 1980s, corn production represented around 35 percent of the total volume of all cereals. In 1960, it scarcely had accounted for 26 percent. Thus, corn's importance surpassed that of both millet and sorghum, which together occupied second place among basic foodstuffs, and perhaps equaled tubers: cassava, yams, and sweet potatoes, which had led the way in the early 1960s (Food and Agriculture Organization, 1968, 1973, 1983; Miracle, 1966: chap. 6). Corn production did not grow at a constant rate. Significant increases in production took place between 1960 and 1975. Since that time, corn production stalled at about 16.5 million tons per year, and no longer kept pace with population growth. Corn production also varied significantly from country to country and from one year to the next, reflecting corn's vulnerability to adverse climatic conditions. Corn production continued to be erratic and some countries in sub-Saharan Africa became regular importers of that grain. Figuring prominently among these were Tanzania, Zambia, Nigeria, which exported petroleum, and Zaire, which exported minerals. Many other countries imported lesser amounts. Other than South Africa, only Zimbabwe figured as a modest corn exporter. Despite any increase in production, increased urban consumption caused supply problems, which were summarily resolved with imports.

Corn production and consumption was not uniformly distributed over the region. In the early years of independence, Marvin Miracle (1966: chaps. 6, 8) ascertained the regions in which corn was most important. Corn was the dominant staple of the majority of the population in a continuous area that included Kenya, Malawi, Zimbabwe, east central Angola, and southern Zambia. Corn also clearly was dominant in some less extensive areas of Tanzania, Benin, Cameroon, and Togo. Without approaching the near exclusivity of the above regions, corn was a leading food among other foodstuffs that routinely made up the diet of widespread areas in Mozambique, Zaire, Uganda, Tanzania, southern Cameroon, the Zande district in Sudan, northern Ivory Coast, and southern Upper Volta. Corn was moderately important in the rest of tropical Africa, and sometimes predominated at certain times of the year. The rise in corn production in the face of the stagnation of production of other basic staples extended the regions where corn was the leading maintenance food, and especially increased corn's relative importance where it was one among several basic foods making up the diet. Corn became an important staple in many African cities that had been subject to recent and explosive growth at rates of up to 5 percent a year. Practically all the corn produced in tropical Africa and even imported corn was earmarked for direct human consumption. Corn's use as fodder or in industry continued to be only marginal.

Corn's growing presence in African diets is related to the dishes in those areas where its consumption is important. Most often, corn is made into a cornmeal mush or a thick paste cooked with water, very similar to Italian polenta, Romanian *mamaliga*, or Spanish *gachas* or *puchas*. Pieces of dough are shaped into small balls. Those portions or mouthfuls of dough are dipped in sauce, or even a stew of meat, fish, insects, and a wide variety of vegetables, with a variety of seasonings. This all serves to help the paste go down and give it some flavor. The dough or paste, which has a neutral taste, is eaten everywhere for every meal and provides the largest part of daily caloric intake. The sauces or stews are what provide variety and a balanced diet. The dough can be made with flour derived from any number of different cereals or tubers. Different flours are often used for the paste in many areas where corn is a staple, reflecting seasonal availability or availability affected by price structure and fluctuation. Obviously, there are preferences that cannot always be accommodated with respect to types of flour. During the 1970s and 1980s, corn's leading role in these diets had been growing and in many areas had become dominant, at times at the expense of more preferred flours. In Africa, as in almost all peasant cultures, nourishment depends on access to a high quality, readily available ingredient. That dietary component is considered to be the true staple, the sustenance of life. It is that food that can prompt the physical sensation and the very idea of being hungry and, conversely, full (Richards, 1939; Miracle, 1966: chap. 8).

Twentieth-century growth in corn production, to the point of making it Africa's chief crop, owed mainly to bringing new lands under cultivation. Productivity in tropical Africa had not grown significantly since the 1960s. The average yield for the period 1960–65 was less than half a ton per acre. Yields for the three-year period 1974–76 rose to 1.2 tons per acre, where they remained at a standstill until the early 1980s. On the other hand, the total acreage under cultivation grew by 50 percent (Food and Agriculture Organization, 1973, 1983). Most producers used the same methods for cultivating corn that had always been used: hoeing corn by hand, burning dry underbrush, freely cultivating other crops between corn rows. Predictably, corn yields did not rise. Miracle's calculations (1966: chap. 1), which put at 80 percent the number of producers who obtained yields in 1960 of just over a ton per acre, appeared to hold true for the late 1980s.

The data showed that in 1960, with the exception of Zimbabwe, the volume of commercially marketed corn did not amount to even one-fourth of total production (Miracle, 1966: 85). This implied that most corn producers directly consumed their own crop. No more up-to-date hard figures are available, but what information there is suggested that it would be possible for the proportion of total production earmarked for the market to rise without jeopardizing self-subsistence. One can infer that such a rise would owe fundamentally to peasant producers taking a larger proportion of their crop to market. The number and importance of commercial or mercantile producers seemed to be at a standstill or even declining. Corn cultivation in the late 1980s continued to be fundamentally a peasant crop. Corn was overwhelmingly grown by family units using traditional methods and played a key role in peasant self-sufficiency. Peasant producers represented the immense majority of productive units that cultivated corn. They produced at least four-fifths of the total amount of corn grown in tropical Africa at that time.

White settlers played a conspicuous role as commercial corn producers. They benefited greatly from the use of the plow as well as other innovations. In some countries these white settlers disappeared or diminished in importance after independence. This was the case, for example, in Mozambique. Four thousand farms formerly belonging to white settlers that had monopolized half of that country's arable land were converted into state-run units. Some of their productive techniques and structure were preserved, even as communal villages were founded on those lands. Things played out differently in Zambia, where some 500 or 600 large private farms had produced nearly half of all corn. Even though nearly half of those farms remained in the hands of European settlers after independence, there was also considerable African ownership of land. Prior to independence in Zimbabwe, some 6,000 private farms belonging to European settlers controlled the most fertile plots of arable land and produced more than half of all corn. Some of these properties were abandoned and there was an attempt to settle African peasants on others. The majority remained under the control of their former owners as did commercial corn production (Dinham and Hines, 1984: chap. 6). While a precise accounting did not exist, the impression remained that the participation of white agriculturists in corn production diminished overall even while it largely was preserved in some countries. What was not preserved were the incentives for growth typically derived from the exclusive nature of advan-

Corn & Dependency

tages and support that colonial governments formerly had conferred on white settlers.

In some cases, African governments tried to compensate for the decline in white agriculture with state-sponsored projects to promote the production of foodstuffs through large-scale agriculture and modern technology. This was the case in Zambia, where an ambitious project was carried out to establish eighteen state farms of almost 50,000 acres each. In Tanzania, the national corn project also attempted large-scale mechanized production by forming blocs of private properties. It was not unusual for these types of projects to have discouraging results when measured against their stated objectives. Typically, foreign capital or foreign aid financed such projects, with the conspicuous intervention of foreign technicians, themselves frequently consultants for multinational corporations. The high operating costs of these undertakings, often a function of imported machinery and consumables, could not be met with the depressed prices of domestic commodities and had to be subsidized. These projects had serious operating problems typically generated by overly authoritarian and bureaucratic administration. The programs chronically fell short of their production and productivity objectives and did not approach the optimistic rates of return projected by technicians. They also ran up against peasant resistance. Peasants insisted on the right to decide production matters, to determine matters of common sense and the associated organization of production (Dinham and Hines, 1984: chaps. 6, 7). The results of projects for the large-scale production of foodstuffs were no more encouraging when measured against other parameters. All economic resources for the development of the production of foodstuffs, both foreign and domestic, were concentrated in these types of projects that excluded the majority of peasant producers from both their hypothetical and actual benefits. In fact, this type of project fomented and heightened the concentration of income and resources and in many cases actually had a negative effect on producers, who remained marginal to their implementation. Furthermore, these costly projects frequently tied up domestic financial resources over long periods of time and were not effective in lessening dietary dependence. On the contrary, they only made it worse.

Private or state-sponsored projects aimed at instituting large-scale modern agriculture for the purpose of feeding tropical Africa apparently failed. This circumstance leads us to take a second look at the central role peasants might play in addressing some of Africa's food issues. The increase in corn production illustrates the strides peasants have already made in that direction. Rising corn production that outpaces population growth raises the possibility of alternatives to larger-scale programs. While the growth in corn production has been encouraging, it has lagged behind demand and became stagnant after 1975. It is self-evident that severe impediments to growth still pertained in the late 1980s. The future of independent Africa is hobbled by something as basic as feeding its own people.

Rising corn production in the 1970s and 1980s stood in stark contrast to decreases, stagnation, or much slower growth rates for the production of other staple foods in tropical Africa. This implied that the rise in corn production was made partially at the expense of other peasant food crops, for which corn sometimes was substituted. In other instances, it implied that peasants planted corn rather than other crops in newly cleared fields. In this case, corn did not expand as a complementary crop. Rather, corn spread in the capacity of a competitor for scarce resources. Peasant producers preferred the American cereal precisely because it offered advantages over other crops. The analysis of those advantages had to take into account the restrictions that came to bear specifically on peasants. Those restrictions included not only natural factors, but also social relations of subordination.

One of the advantages corn enjoyed was its effortless adaptation to a large range of natural settings, including poor soils, rugged terrain, or high altitudes. Such inauspicious conditions had been the norm for African peasants since the late 1960s. The confinement of peasant producers to small areas, with the least fertile soils, was a result of the resettlement of the population under colonial rule in order to favor the expansion of agribusiness or commercial agriculture. Agrarian inequity endured and in many cases became more pronounced in newly independent African nations. Prevailing territorial restrictions made corn even more vulnerable to unfavorable climatic phenomena. Unstable yields came to characterize corn expansion. The risks to which the corn cultivation was subject—its cultivation on marginal lands or on depleted soils—were fewer than those to which other competitive plants were subject, but were growing and entailed increased uncertainty for their producers. Because of such agricultural limitations, high risk and uncertainty became norms that producers working marginal lands had to take into account when defining their productive strategies.

Figures from Miracle's book (1966: chap. 11) are the basis for the comparison of corn with other staple crops, although I alone am responsible for any analysis of those figures as they pertain to peasant societies. Corn yields in West Africa of around 0.36 tons per acre were equal to yields for rice and were 15 percent higher than yields for sorghum and millet grown under similar conditions. Corn's higher caloric value gave it a slight edge when yields were

Corn & Dependency

evaluated in terms of nutritional content. Corn had a slight advantage in the grasslands of the savannas, the most extensive open region in tropical Africa. By way of contrast, rice had an advantage in the tropical jungle and sorghum thrived in the drylands bordering the deserts. On the other hand, bulk corn yields were considerably less than those of tubers: 3.6 tons per acre for cassava, 1.8 for sweet potatoes, and 1.3 for yams. The low caloric content of tubers, pound for pound, meant that the actual differential in yields between corn and tubers was much less if calculated in terms of calories per acre. Even so, cassava produced 3.5 calories, yams 2.1 calories, and sweet potatoes 1.5 calories more per acre for every calorie of corn. In general, tuber cultivation required more moisture than corn cultivation, but in many areas the crops coexisted and were competitive. This circumstance meant that neither gross yields nor nutritional yields alone were enough to explain the preference for corn.

The number of work days required to cultivate and harvest an acre of corn could vary from twenty to eighty. Something less than forty work days per acre of corn was the most common figure in the case studies compiled by Marvin Miracle (1966). Invariably corn required much less work than rice, which was labor intensive. Corn also was less labor intensive than millet. The comparison with sorghum was less telling and at times sorghum had a slight advantage over corn. The same was true of cassava, although in most cases corn required less work. If all the tasks necessary after harvest to transform the crop into flour were taken into account, all the cereals required a similar amount of labor. However, it should not be forgotten that corn also was eaten green right off the cob, with no additional effort needed. This circumstance did not apply to the other grains. The transformation of tubers into flour demanded a much greater investment in labor than that required by cereals, almost twice as much. Actual corn yields aside, the total amount of work needed to raise and transform corn was the lowest of all the staple crops, perhaps with the exception of sorghum. This could have enormous importance for peasant producers on two different levels. Corn's low labor input was a great advantage for productive units that could not assign their entire workforce to agricultural production. This applied especially to those cases in which a part of the peasant workforce, almost always young men, had to migrate or contract out locally for a salary in order to obtain or supplement monetary income. In those many areas in which agricultural work, with the exception of clearing, was carried out by women and children, such a criterion could be crucial. On the other hand, corn's low labor requirements implied lower losses in the case of natural disaster. This was very important in those areas where agriculture was practiced under extremely risky conditions.

Corn stands out among the staple foods of tropical Africa as the earliest maturing crop, the crop with the shortest growing period. Corn requires between two and five months to mature, the norm being somewhere around four months. Raised under comparable conditions, corn matures a full month before sorghum, millet, or rice. Here it is important to add that corn, as green corn, can be eaten up to a month before fully maturing, when scarcity of food is most sharply felt in the period immediately preceding the harvest. Corn's short growth cycle is even more marked in comparison with tubers. Corn can be harvested at least five to eighteen months before tubers. Two or even three corn crops easily can be harvested over the same length of time it takes for one cassava crop to mature. Certain varieties of cassava can take up to two years to reach maturity. In this sense, corn surpasses or cancels out any difference in nutritional yields, although multiple corn harvests imply higher labor inputs. Besides corn's availability for consumption at the most critical time of year, early maturing corn varieties have several advantages for peasant producers. The most important of these is the short turnaround time necessary to obtain foodstuffs or money income. That is a very important criterion when there are no food reserves or storage, when life is lived day to day, as it is by the majority of peasant producers. This criterion can be definitive for newly formed productive units as they are incorporated into the agricultural work cycle. It is important that such units survive, as they probably are the most common vehicle for the expansion of the corn. Corn also represents a means to reduce risks or more rapidly recover from natural disasters. The less time devoted to corn cultivation, the sooner labor is freed up and able to return to the sale of its labor power in other pursuits.

Corn's high performance is attributable to its adaptability, its high yields relative to other cereals, its low labor inputs, and its short growth cycle. All these factors go far in explaining why peasant producers prefer corn over other crops. The combination of these factors acquires special significance when producers are severely restricted in their access to resources, restricted in the amount of effort that they can devote to agriculture, restricted in the amount of time that they can wait to have food or obtain income, and restricted by conditions of poverty and lack of aid. Producers cannot wait two years just because cassava promises a bigger harvest with less work. They cannot tolerate the lower bulk yields of sorghum or millet just because they are the main ingredient in some of their favorite foods. Corn's advantages for the peasant become clear-cut in circumstances of poverty, of urgency, and of risk.

On another level, corn as a commodity also has advantages for peasant producers. Corn has the best-consolidated, farthest-reaching, and longest-

Corn & Dependency

standing markets. Corn's compact nature, its easy handling, transport, and storage, almost always result in open-ended demand, although not always at attractive prices. The sale of corn is a sure thing. Its movement between the home, where it is used for food, and the marketplace, where it is turned into money, is relatively effortless and fluid. This fluidity gives peasants limited options. Ideally, it allows peasants to select the best remunerated work. The options for peasants looking for work and negotiating salaries, however, are few. Neither are peasants able to exercise much control in setting prices for their agricultural goods. Their options come down to seeking out the least detrimental balance of profit and loss.

Corn flour is one of the cheapest foods per thousand calories found in urban African markets. Imported wheat bread or subsidized rice is occasionally cheaper, but the flours of the other African staples almost never are. Corn's low market price is the result of many and complex pressures. Governments often act as monopolistic buyers and set an official price. Low prices aim to favor urban consumers, whose voice has more resonance and political power than that of the peasant. At times these prices are below cost in order to import subsidized corn or its equivalent from the United States. At other times, prices are determined by the availability of financial resources, almost always scarce. Tanzania made one of the most original attempts to achieve autonomous development by raising the official price for corn paid to producers in order to combat growing dependence on foreign imports. Peasants increased the volume of their commodities destined for the marketplace. Lack of financial resources limited what purchases the government could make, and the merchants and middlemen took advantage of the situation in order to severely lower corn prices (Dinham and Hines, 1984: chap. 5). Backed by large capital, they also pushed for the lower prices paid to peasants. The interests of peasant producers were subordinated to the interests of the urban sector and foreigners, key players in many of the newly independent states. The undermining of rural agricultural markets was for their benefit. Even with all corn's advantages, the low price of corn meant that peasant producers still were compelled to suffer low rates of return on their labor.

In that framework, any strides in peasant corn production could be interpreted as a defensive adaptation to increasingly deteriorating conditions rather than a positive response to economic stimuli and incentives. Corn production was less risky and could preserve peasant autonomy in a limited way. Turning to corn also held out the prospect of more work for less money. Increases in corn production at the expense of other food crops suggested that any advantage associated with corn served to exploit producers even further. If this hypothesis was correct, slowing corn production signaled that such an alternative already had reached its logical limits. Advantages already had lost their elasticity; they already had provided whatever edge they could. Peasants did not resort to expanding corn cultivation as a means to resist foreign pressures, which were increasingly onerous. Meaningful agrarian reform that broke with regional restrictions that weighed so heavily on peasant producers could provide the only catalyst for growth in the production of food in tropical Africa. It was imperative that measures be adopted to correct the unequal relations between rural peasant producers and national and international centers of power. Until that day comes, peasants will resist by using corn and extreme poverty as a shield to protect themselves. Corn and other basic staple crops will permit them to survive, even if only in the most precarious of circumstances. Anything in order to endure.

: 8 :

CORN IN EUROPE: AN ELUSIVE TRAIL

Corn was an important commodity in late-twentieth-century Europe. Corn ranked third among cereals, just behind wheat and barley, and accounted for a little more than 20 percent of all European grain production. Not including the former Soviet Union, Europe produced more than 55 million tons of corn annually during the 1980s, accounting for one-eighth of production world-wide. Most of this corn was earmarked for animal feed. Corn was cultivated everywhere in Europe except for the Scandinavian countries. Besides the former Soviet Union, three European countries ranked among the world's ten top corn producers: Yugoslavia, Romania, and France, which produced more than half of all corn grown in Europe. In four countries—Hungary, Portugal, Romania, and Yugoslavia—corn production was higher than wheat produc-tion and in two—Austria and Bulgaria—corn and wheat were produced in equal amounts. The Balkan countries and the countries of the Danube River basin made up one of the world's chief productive regions. In France, Italy, and Portugal, corn was one of the leading agricultural products. Average corn yields were 20 percent higher than both wheat yields and yields for all cereals combined, and the gap continued to widen. The expanding nature of Euro-pean corn cultivation remained a powerful and growing current.

There was a wide array of different kinds of evidence pertaining to corn's introduction and diffusion in Europe, the wellspring of that fast-flowing current. After more than five hundred years, much of this evidence had disappeared or been compromised and faded away. Other evidence was ob-scured, waiting to be rediscovered. What evidence there was of corn's migra-tion to Europe was discontinuous, incomplete, and at times confusing to the point of raising more questions than answers. Our poor understanding of the adoption of American plants and their effect on Europe cannot be explained away solely on the basis of the various problems posed by our sources. The range of available material was wide. What was lacking was much of an interest in the topic to begin with. There were some exceptions, of course, including some outstanding monographs. But the material did not form a school of thought or a well-defined area of specialty. The impoverished nature of the information available on American plants in Europe was all the more remarkable viewed within the wider context of European agricultural history. That body of literature was the most extensive and well documented in the world, although not always the most free-thinking, and corn was only begrudgingly allowed a place in it.

Those attempting to reconstruct the history of corn's European migration frequently and consistently cited entries on corn from early New World chronicles. Nicolò Syllacio wrote a pamphlet published in Pavia, Italy, in December 1494, which was the very first European publication to acknowledge the existence of the American cereal. Syllacio described corn, although he did not name it. Corn was just one among many New World novelties (Weatherwax, 1954: 31–33). Peter Martyr described corn in his *Decades of the New World*, although he too declined to name it. The first edition of his first volume appeared in 1511. Martyr's was the first book-length publication in Europe to refer to the corn plant. Martyr added a name for corn to its description in the book's second edition, published in 1516. That description verified that seed corn was plentiful among Andalucians and Milanese. This gave rise to several different interpretations, almost all tending to support the Old World origin of corn or its introduction well before Columbian contact.

Gonzalo Fernández de Oviedo came to America in 1514 and died here in 1557. He wrote the *Sumario de la natural historia de las Indias*, published in 1526. Oviedo published the first part of his *Historia general y natural de las Indias* in 1537. It was reprinted in Salamanca in 1547 and in Valladolid in 1557. The book was translated into French in 1555 and into Italian in 1556. Oviedo's *Historia general* was a true publishing success story. In both of Oviedo's works there were a number of chapters that described America's natural setting. Oviedo dedicated one chapter specifically to a description of corn, including its cultivation and use. This was the first rigorous, classic, and scientific description of corn (Weatherwax, 1954: 34–47; Gerbi, 1985). Other New World chroniclers, including Las Casas, Acosta, Francisco López de Gómara, and Pedro de Cieza de León, published works in Europe over the course of the sixteenth century. These early chroniclers added descriptive material on corn and corn's significance in America, but said little about the arrival of the plant on European soil (Mesa Bernal, 1995: chap. 9).

Many other sixteenth-century New World chroniclers never lived to see their works published. Some were published posthumously. Among these, Francisco Hernandez merited special mention. Hernandez visited Mexico between 1570 and 1577, commissioned by the king of Spain to make a compilation of indigenous medical knowledge. Hernandez prepared a lavishly illustrated work on Mexican flora and fauna, at times making use of the reports of Friar Bernardino de Sahugún, another one of the luckless sixteenth-century scientists whose works were not published until much later. Hernandez's original work, submitted to his patron, was destroyed in a fire at the monastery in Escorial, Spain, without ever having been published. But rumors began to circulate about the unpublished work, and other writers published excerpts or bits of information from drafts and summaries. Such was the case of Francisco Ximenez's *Cuatro libros de la naturaleza y virtudes medicinales de las plantas y animales de la Nueva España*, published in Mexico in 1608. Ximenez based his book on the Hernandez manuscripts. Hernandez had a late influence on European botanists and naturalists with the publication of a summary of his original manuscript in Rome in 1651. This was known as the book's Roman edition or *El tesoro* by virtue of its Latin title, *Rerum medicarum novae Hispaniae thesaurus*, and was lovingly corrected and revised by the devoted members of the Italian Accademia dei Lincei. In 1790 Hernandez's drafts, located in Madrid, were published in their unfinished form, with the title *Obra*, also known as the Madrilenian edition. Hernandez's *Obra completa* finally was published in Mexico beginning in 1960. The troubled tale of Hernandez's work illustrated the delay and difficulty with which firsthand precise knowledge about American nature became part of the European bibliographic tradition.

Herbals were very popular in sixteenth-century Europe. These were descriptive illustrated catalogues of useful plants, especially plants with medicinal purposes, and usually included information on the history, properties, and practical applications of individual plants. This type of treatise had been produced since ancient times and manuscripts were copied during the entire Middle Ages. With the use of woodcuts and with the introduction of the printing press in the second half of the fifteenth century, those catalogues became more or less affordable books and enjoyed a great deal of demand. New treatises were written in the sixteenth century that respected classic and medieval traditions, but that also depicted new plants as part of the standing inventories (Blunt and Raphael, 1979; Weatherwax, 1954: 34–47; Finan, 1950; Messedaglia, 1927: chaps. 9–12; Reed, 1942: 57–80).

Herbals were useful in following corn's introduction in Europe. In general, their authors did not have a direct or at times even a minimal knowledge about America as opposed to New World chroniclers. Because of that, it was assumed that the inclusion of American plants in their books constituted a sort of record of the plants' diffusion on the continent. This supposition carried with it a number of caveats. The most serious of these was that entries in the herbals generally did not distinguish between corn's diffusion as an ornamental as opposed to an agricultural crop. Corn apparently spread through a good part of sixteenth-century Europe as a garden plant, a botanical curiosity. Corn was not widely disseminated as a staple or a commercial crop at that time. Besides, authors freely borrowed texts and illustrations and, with these, many myths and biases were transmitted indiscriminately along with rigorous knowledge.

Corn first appeared in herbariums in 1539. The German naturalist Jerome Bock called it *welschen corn* or "strange grain." Bock's cost-conscious editor refused to pay for woodcuts, and Bock was forced simply to describe plants using colloquial language and make do without illustrations. Bock both admired and was astonished by the corn plant. In spite of any misgivings he might have had about corn, Bock ultimately recommended an infusion of its leaves as a remedy against erysipelas. In the 1542 herbal by Leonhard Fuchs, another German naturalist, corn figured again, this time with the name "Turkie wheate." A beautiful illustration accompanied Fuchs's description of corn. Other herbals published at later dates reproduced this illustration. Fuchs's text confirmed that by the mid-sixteenth century corn already was growing in gardens all over Germany. Corn joined the ranks of the most popular herbals after being written about by these prominent naturalists. The herbal by the Italian naturalist Petrus Matthiolus, originally published in 1565, was the first to emphatically deny the Oriental origin of corn. Matthiolus was familiar with the work of Oviedo, translated into Italian by Giovanni Battista Ramusio, and the work of other New World chroniclers, which apparently was not the case with his predecessors. The academic debate that would last several centuries had been unleashed (Finan, 1950; Blunt and Raphael, 1979: 132–52).

The illustration included in Fuchs's 1542 herbal and the illustration attributed to Oviedo, which did not appear in the original edition but in the subsequent Italian translation of 1554, were considered to be the first graphic illustrations of corn to appear in Europe. Roland Portères (1967) found an earlier illustration in the frescos of the Italian artist Rafael Sanzio, modestly nicknamed "El Divino." Sanzio's frescos at the Villa Farnesini were painted around the theme of the "History of Love and Psyche," and dated from around 1516, only a few years before the artist's death. Three ears of corn, among other edible Old World plants, appeared in the fresco's upper frieze in a scene in which Venus revealed humanity to Psyche. Luigi Messedaglia (1927) found corn adorning the columns of the Duke of Venice's palace, constructed around 1550. Orlando Ribeiro found stylized corn in Portugal as a decorative element in Manuelian architecture dating from the first quarter of the sixteenth century. The purpose of this digression is to suggest that other data on corn's introduction in Europe, such as iconographic evidence, have received very little attention as opposed to the bibliographic evidence. The excessive reliance on esoteric written records narrowed knowledge and discourse on the impact of New World flora on Europe.

John Finan (1950) compiled a useful survey of corn in European herbals. He tried to demonstrate that two different types of corn were known in Europe in the sixteenth and seventeenth centuries. Presumably, one was acclimated from well before the time of contact. His work demonstrated the degree to which opinion was divided among naturalists who penned the herbals. Half of them sustained that corn originated in the Old World. The majority of authors who adhered to this position came from Germany, the Low Countries, and England. They were familiar with corn only as an ornamental plant. Corn cultivation and corn's use as a food were not widespread at those latitudes. The handful of Spanish and Italian naturalists consulted by Finan favored the American origin of the plant. They were from countries where corn had already become part of the local agricultural scene and where producers directly consumed the corn they grew.

Corn's introduction did not inspire unbridled enthusiasm among European naturalists. John Gerarde, English author of an herbal published in 1597, summed up the prevailing opinion of contemporary European naturalists when he wrote, "Turkey wheate . . . doth nourish far lesse than either Wheate, Rie, Barly or Oates. The bread which is made thereof is . . . hard and dry as bisket is . . . for which cause it is of hard digestion, and yieldeth to the body little or no nourishment, it slowly descendeth and bindeth the belly. . . . We have as yet no certaine proofe or experiences concerning the vertues of this kinde of Corne, although the barbarous Indians which know no better, are constrained to make a vertue of necessitie, and think it a good food; whereas we may easily judge that it nourisheth but little, and is of hard and evil digestion, a more convenient foode for swine than for men" (quoted in Finan, 1950: 148).

What type of experiments might the good Mr. Gerarde have carried out in the privacy of his own garden that led him to draw such a hostile conclusion? Gerarde's negative opinion of corn prevailed over the more commonsense observations made by naturalists such as John Parkinson. Parkinson too was English and he published his herbal in 1640. In it he wrote: "Many doe condemne this Maiz . . . for wee finde both the Indians and the Christians of all Nations that feede thereon, are nourished thereby in as good manner no doubt as if they fed on Wheate in the same manner" (quoted in Finan, 1950:

148). This last excerpt suggested that corn was a common food for both man and beast in the mid-seventeenth century. The bias against corn, however, was already entrenched and has not been completely dispelled to this day.

The publication of herbals dropped off in the last quarter of the seventeenth century, even though their popular demand probably continued to rise. Corn and other American plants almost disappeared from the pages of books for nearly a century. They reappeared in the second half of the eighteenth century in two types of works: systematic botanical treatises, like that of Linnaeus, and agronomy manuals. Both treated corn in a matter-of-fact way, just like any other plant in the European plant repertoire. The discovery of America had already been assimilated, although European assessments of that event continued to be equivocal (Elliot, 1970). The fact that eighteenth-century texts included corn as a matter of course did not imply that the speculation about corn's origin had been cleared up or forgotten. It was the era of Georges-Louis Buffon and Cornelius De Pauw and, a little later, Hegel. All these intellectuals preached the natural inferiority of America. America was undeveloped, decadent, and corrupt. One could infer that America's subordination to Europe was a natural and logical state of affairs (Gerbi, 1973). Linear evolution and Eurocentrism arose as elements of the dominant intellectual paradigm. With great satisfaction, European intellectuals deemed European nature and civilizations to be the acme of the evolutionary process. There was little sympathy for a position positing corn as a product originating in the New World.

Works on agronomy sought to extend and improve the cultivation of corn and other American plants. Their cultivation would contribute to meeting the growing demand for food that resulted from sharp demographic growth and urbanization. Many agronomy books were products of meetings convened by scientific societies. The membership of these societies was worried about the well-being of the population in the second half of the eighteenth century. The treatise on corn that had most resonance was that of Antoine-Augustin Parmentier. Parmentier was a French agronomist who a decade before had published a monograph on the potato. Parmentier's book on corn was entitled *Memoire sur le mayz* and was published in 1785 by the Academia Real de Burdeos. Parmentier was so well known that some attributed his book with the diffusion of corn throughout Europe. To conclude that a book had more influence on the spread of corn production in Europe than actual circumstances affecting everyday life was, of course, nothing more than idle speculation. Parmentier wrote his monograph in order to promote the most effective methods to plant, care for, harvest, store, and use corn. He supposedly based the book on the experience of real farmers and it had great influence, establishing a model for later agronomic monographs and treatises.One of the book's flaws was that the text was a seamless stream of commentary and observation. It was virtually impossible to clearly distinguish between descriptive passages on traditional practices in corn cultivation and passages containing new, innovative practices proposed by the author. That ambiguity was common to many eighteenth- and nineteenth-century works on agronomy.

Some treatises on agronomy began to appear by the sixteenth and seventeenth centuries. Gabriel Alonso de Herrera, a Spaniard, wrote *Agricultura general*, which would become very well known. The first edition appeared in 1513, but de Herrera made no reference to corn. Corn was acknowledged in some seventeenth century works, but there were no recommendations concerning its cultivation. While agronomy books were no longer novelties by the eighteenth century, it was not until the mid-1700s that the situation definitively changed. It was then that enlightened elites, gentlemen farmers, began to show an interest in agriculture (Fussell, 1972). The work of Antonio Zanoni, *Dell' agricultera, dell' arti, e del commercio*, published in Venice in 1765, made practical recommendations on corn cultivation. The *Encyclopédie*, coordinated by Denis Diderot and Jean Le Rond d'Alembert and published between 1751 and 1763, included two articles on corn with recommendations for its use and cultivation. One undated work by an unknown author was perhaps the first monograph wholly dedicated to corn cultivation. It was published in Italian in Berlin with the title *Dissertaziones della coltura e utilita del grano-turco*, very probably even before the appearance of the monograph by Parmentier. In 1788 Gaetano Harasti da Buda, a Franciscan friar originally from Hungary, sponsored by the Accademia Agraria di Vicenza, wrote the pamphlet *Della coltivaziones del maiz*. Probably the most important monograph on corn identified with this trend in Europe owed to Matthieu Bonafous, *Histoire naturelle, agricole et économique du maïs*. It was published in Paris in 1836, and was almost a modern work in terms of its organization, complete with critical apparatus and extraordinary illustrations. This book, which once again raised the possibility of the Asiatic origin of corn, marked the end of an era. After this time, debates surrounding corn would revolve completely around pellagra, the curse of corn.

In the twentieth century, historians aggressively began to pursue references to corn in archival documents. There were very few sources written by Europeans. In the first half of the twentieth century Luigi Messedaglia stood out, writing two voluminous books about corn in Italy. In those, he relied on the usual bibliographic sources and supplemented those using other types of documentation. By the mid-twentieth century, specialists in human geography who had been influenced by diffusionism, especially in Germany and Portugal, investigated and wrote about the migration of corn and other plants. In the second half of the twentieth century, quite a few other authorities had been added to the list, almost all linked to the French journal *Annales* and associated with Fernand Braudel. All of these were influenced by U.S. authors, especially by biologists, who had reopened the old debate about corn's origin and dispersion. Much archival research remained to be done. It still had not begun in many countries and many of the questions already posed on the adoption of corn cultivation went unanswered. The topic, while it had received a certain degree of attention, was not an especially trendy one.

It is possible to outline the course of corn's spread in Europe in a schematic way using the little information we have, even allowing for any gaps and lapses in continuity. We do not know with any certainty who the growers and propagators of New World seed corn were, although we can imagine their astonishment at the appearance of this strange, previously unknown plant. Most authors agreed that the propagation of seed corn took place in Spain. Oviedo claimed to have seen corn stalks ten hands high in the vicinity of Avila, Castille, sometime before 1530. Seed corn production probably was concentrated in the irrigated lands of Andalucia in the first two decades of the sixteenth century. Efforts did not go unrewarded, with corn plants effortlessly producing plentiful amounts of seed. We know much less about who might have promoted the diffusion of corn as an ornamental and kitchen garden plant, an informal process in which a very limited number of seeds were scattered in all directions. Naturalists probably played an important role in this process. Thanks to them, we know that ornamental corn already was widely known throughout Europe by the beginning of the seventeenth century. Also thanks to them, we know that corn was planted in botanical gardens and that it was tended in herbariums as part of botanical collections. At that time, the acclimatization of New World plants and plants from other regions was seen as one of the primary missions of botanical gardens (Reed, 1942: 113–25).

When and where corn proved itself as an agricultural crop also remain unknown. Indirect references seem to suggest that this process took place in the irrigated lands of Andalucia in the Guadalquivir River valley in the early years of the sixteenth century. Unfortunately, we do not have at our disposal any work tracing corn in Andalucia. Ironically, corn ultimately did not attain or sustain the same level of importance there that it came to have elsewhere in the Iberian peninsula.

Corn was an irrigated kitchen garden crop in Valencia in the last third of the sixteenth century. By the eighteenth century, corn totally had displaced millet and panic grass, the traditional cereals of summer. Corn also deprived these crops of their very names, in that the term "corn" came to be applied to all of these cereal grasses collectively. At that time, corn already had been incorporated into normal crop rotations together with wheat, hemp, and beans. Corn was eaten together with other cereals as part of a traditional diet. Nevertheless, because of corn's lower price, its consumption was more prevalent among the poor. In years of scarcity, such as 1741, corn simply substituted for wheat and barley. Corn still was eaten in the late nineteenth century, although in lesser amounts. On the other hand, the cultivation of corn for fodder was on the rise. Corn was one of the most important crops in the history of irrigated kitchen gardens of the Spanish Levant, one of Europe's richest and most varied agricultural regions, where Arabian irrigation techniques were first introduced and ultimately preserved.

Corn was very important in northern Spain. In Asturias, an old belief attributed corn's introduction to Gonzalo Mendez de Cancio, who would be governor of Florida in the New World between 1597 and 1603. It was said that it was Mendez de Cancio who brought seed corn from America to sow in Vega de Bria and that one of the large chests used for that purpose had survived. In the first half of the seventeenth century, other testimonials mentioned corn's introduction in Asturias. In the eighteenth century, the importance of the corn crop and its use as food already was significant. Following received wisdom, José Gómez Tabanera (1973) suggested that corn was introduced into Galicia and the Basque region from Asturias. Other sources cited by Jaime Carrera Pujal (1947) pointed to corn cultivation originating in Guipúzcoa, in the Basque region in 1576. In 1630 corn production remained insignificant, but by the eighteenth century corn had surpassed wheat and was the leading staple crop. Rural residents thought that corn was more nutritionally sound than wheat. In eighteenth-century Galicia, corn already was a leading staple crop, second only to rye. Eighty percent of Galicia's population was subsistence based, with very little money in circulation, and corn was one of the region's fundamental staple crops (García-Lombardero, 1974).

Humid northern Spain was densely populated, but that population was dispersed in small villages and rural hamlets. Subsistence agriculture was the norm, and corn became one of the building blocks of the peasant economy. Corn served equally well for feeding humans and those domestic animals that formed an integral part of the economy. We do not know if corn was introduced into northern Spain as an irrigated crop, although this probably was the case in Portugal. In any event, by the eighteenth century corn already had become a seasonal crop in northern Spain.

The scarcity of works on corn's introduction to Spain contrasts with the relative abundance of those at our disposal on Portugal (Ribeiro, 1955; Godinho, 1984; Dias, et al., 1961: 235–53; Mauro, 1969: 296ff.). There was general agreement that corn had to have been adopted between 1515 and 1525 in Coimbra. In 1533, corn prices already were quoted on local commodity exchanges. Corn was priced 20 percent cheaper than wheat, but higher than rye, barley, millet, and panic grass. By the middle of the seventeenth century corn already was a common food in the provinces of Minho and Beira. By the eighteenth century corn abounded in all of northeast Portugal, the country's most densely populated region. The widespread use and cultivation of corn probably extended from Portugal toward Galicia.

Corn was introduced to Portugal as an irrigated crop in fertile river valleys. The irrigation of terraces and platforms had been practiced since before Roman times. Prior to the introduction of corn, irrigation had only been used to water pasture lands during the winter months. In the winter, both good agricultural land and impoverished soils were devoted to raising cereals such as wheat, rye, and barley using only dry farming techniques. The low yields of summer cereals such as millet and panic grass meant that for four or five months of the year inhabitants of the northeast were forced to live on nothing more than chestnuts. The introduction of corn brought the transformation of irrigated pasture lands, formerly idle during the summer, into cultivated fields. High yielding corn enabled the people to face the lean months without hunger. The new and intense utilization of the soil through field-fallow rotation brought about the expansion of irrigation and the construction of new terraces and platforms.

The high yields for irrigated corn, higher than that of other cereals, caused corn's price to fall and eased seasonal fluctuations in supply. In less than a century after its introduction, corn cultivation swept through the Atlantic lowlands. In the two centuries that followed, corn spread to the mountains, where it also became a dry farmed crop, displacing millet and rye and contributing to the decline of chestnut trees. Corn's introduction in Portugal transformed agriculture, stock raising, settlement patterns, and the rural landscape itself. Corn furthered demographic growth and rural migration. In the nineteenth century and the early years of the twentieth, corn monopolized more acreage than any other cereal grown in Portugal. In the twentieth century a costly government policy promoting wheat cultivation changed that situation and caused severe disruptions in the ecology and economy of the rural countryside.

Corn cultivation extended from Spain into neighboring France in the mid-sixteenth century. Corn appeared in the provinces bordering the Pyrenees, especially on the outskirts of Bayonne and in the region of the Landes, in the southwest, and along the Atlantic coast. Another nucleus of corn's early introduction extended along the Mediterranean rim, in the French-Catalonian zone, from where it spread toward the province of Languedoc. In the seventeenth century corn got a foothold in the Midi, in the southeast, where its cultivation overshadowed that of wheat and the use of corn as food became widespread. By the eighteenth century corn figured as a common agricultural crop in the southern half of France, including Guyenne, Languedoc, Navarre, Borgoña, and Franco Condado. In the late eighteenth century, Arthur Young, a traveling English agronomist, drew a line across a map of France at a forty-five degree angle. It extended from the mouth of the Garonne, to the north of the city of Burgundy, to the Rhine in the northeast, to the north of Strasbourg. To the south of that line corn was grown and, according to Young, brought about the disappearance of fallow or untilled lands. We do not know if corn also was introduced as an irrigated crop in France, but its expansion in the eighteenth century suggested its widespread cultivation using dry farming techniques in rotation with other winter cereal crops. Except in regions like the Landes, corn was part of the French diet, but that diet did not become dependent on the American cereal (Hemardinquer, 1963; Roe, 1973: 46–54).

On the Italian peninsula, the cultivation of corn as an agricultural crop as opposed to an ornamental plant began around 1550 on the Venetian plain. The geographical diffusion of corn was discontinuous. Corn's sudden appearance in Venice was inconsistent with diffusionist notions that the plant steadily expanded out from the Iberian peninsula. The clear implication was that the reproduction of seed corn was carried out in Italy and Spain simultaneously and independently. That was not surprising given the fact that, for complex reasons, contact with the New World had a tremendously stimulating intellectual impact on the Venetians, apparently a greater impact than on the Spaniards themselves. By 1571, according to Matthiolus, corn already was eaten in Venice, and at the end of the sixteenth century corn flour was mixed with wheat flour and flours from other cereals to make bread. In the Venetian district of Udine, there were records of the market price for corn dating from 1586. This allowed for the reconstruction of the longest series of prices for that American grain in Europe. In the first half of the seventeenth century corn prices in Udine oscillated between 20 to 50 percent below prices for wheat (Braudel and Spooner, 1967). In the seventeenth century corn cultivation became increasingly important in Venice and spread toward Lombardy, Emilia, and Verona. In those regions, which share the Lombard plain—the country's leading agricultural region—corn spread primarily as an irrigated crop. The introduction of corn probably was one of the reasons behind the expansion of irrigation works in this, Italy's largest plain and its best agricultural land.

Probably about this same time, the first half of the seventeenth century, corn cultivation spread through the valleys of the Alpine piedmont until it reached Austria and the Balkans. In the eighteenth century corn cultivation spread to all of north central Italy and its cultivation displaced that of other summer cereals—sorghum, millet, and panic grasses—in almost the entire peninsula. Corn was used chiefly, almost exclusively, for food. Polenta—pasta or corn dough—had become the prevailing and in many cases the sole nourishment of poor peasants in Italy's north central region. Traditional crop-fallow rotations had been disrupted. The introduction of corn allowed multicrop rotations, alternating a winter cereal and summer corn. Multicropping radically modified the agrarian landscape of the north central portion of the Italian peninsula, Italy's richest and most densely populated region (Messedaglia, 1927; Sereni, 1997).

There was no evidence of corn cultivation in the Balkans before the seventeenth century. The oldest reference to corn placed it in Croatia near the Venetian Republic in 1611. The crop probably extended outward from there. Corn also was introduced near Constantinople about this same time. Corn's dispersal throughout the Balkans fanned outward from these two early primary centers and from other secondary centers, like the Peloponnesus and Dalmatia. Corn progressively penetrated the entire region, from the coasts toward the interior following the course of rivers and river valleys. By the eighteenth century corn cultivation was widespread. Montenegrins adopted corn as their dietary mainstay. In other places, such as Albania, corn was basically a crop for export to the West. The Balkans were at that time under Turkish rule and it seemed that the introduction of corn owed to the interests of Turkish large landowners and overlords in promoting corn as an export crop. In Bosnia, Turkish beys were even known to thrash peasants in order to force them to sow corn. Despite such coercive measures or perhaps because of them, corn exports from the Balkans increased over the course of the entire eighteenth century.

Corn cultivation was introduced in Romania in the seventeenth century. Corn was in Bulgaria and probably also in Hungary by the eighteenth century.

Corn in Europe

Corn fanned outward from the Balkans and probably too from Austria to the Danube River basin. The expansion of the crop in that region perhaps responded to mechanisms similar to those used in the Balkans, although some of these new overlords were Christians. By the nineteenth century Balkan countries and the countries of the Danube River basin had become important corn producers and exporters. Consequently, corn had been successfully adopted as a popular foodstuff in that region, displacing sorghum and millet. The most common explanation for this displacement was found in corn's high yield (Stoianovich, 1951, 1953).

Corn was introduced into Russia and the surrounding region in the early seventeenth century, probably spreading out from the Balkans during the period they were under Turkish influence. Corn was cultivated by non-Slavic peoples in the Carpathians and Caucasus borderlands. Corn substituted for millet as the staple grain in Bessarabia and Imeritia in the central Caucasus. Toward the end of the eighteenth century, corn spread among Slavic peoples of the Ukraine, the Kuban lowlands, and Georgia in the southern part of European Russia. Corn spread quickly through southeastern Russia after the incorporation of Bessarabia by Russia in 1812 (Kupzow, 1967–68; Anderson, 1967).

On the eve of the nineteenth century, at the height of the revolutionary era, corn had a firm foothold in European agriculture. Corn was everywhere in southern Europe, from the Black Sea in the east to the Straits of Gibraltar in the west. Corn's strongest presence was along the Mediterranean rim, under irrigated agriculture. Corn cultivation was no longer limited by irrigation, however, and corn's penetration into the interior of the continent already clearly was underway. Corn had spread throughout the Atlantic littoral, from Portugal to France, where the influence of the Gulf Stream moderated temperatures and produced moisture, and continued its push into the interior. The line traced by Young had long been accepted as dogma and it was firmly believed that corn simply did not prosper anywhere north of Nancy, France, due to low temperatures and short summers.

Corn had been adopted almost exclusively for use as food in those places where agriculture and livestock were dissociated, such as the Mediterranean. Corn was introduced into European diets in two different forms. One form was that of flour. Corn flour was mixed with flours made from Old World cereals in order to produce low-quality breads. It was quite a common practice to mix flours, and most breads were made that way. White bread made with refined wheat flour was not the rule, even among the rich. In some parts of Europe bread was made using only corn flour, although this practice was not widespread. The other way to consume corn was as corn flour dough cooked in water. These pastes had been common all over Europe since ancient times. In areas where corn was cultivated, paste made from corn flour was cheap, abundant, and extremely widespread as a peasant staple. Corn paste took the place of pastes prepared from other cereals. Corn paste also was a common food among the urban poor because of its low price. In most corn-producing regions, a part of the harvest was reserved for the self-sufficiency of producers. The rest was destined for market. Rural overlords from the Balkans to northern Spain appropriated that part of the corn harvest that corresponded to them as owners of the land. They actively encouraged the burgeoning and increasingly important corn market, which already extended far beyond their national borders.

In other parts of Europe, like the rural areas of the Atlantic littoral, people and their livestock were inseparable. With little or no circulation of money, corn provided nourishment for both people and animals. The virtues of corn as forage already had been widely recognized, utilizing not just the kernels but the leaves, tassels, and silks as well. Corn's large-scale use as forage, however, still was not widespread in the eighteenth century. Corn cultivation exclusively for forage spread quickly in the nineteenth and, especially, the twentieth century. Typically, the ear was not allowed to mature in corn cultivated exclusively for silage. It was for this purpose that early maturing varieties that were not especially attractive for human consumption were introduced. Since the maturation of the ear was not an issue in corn for silage and the growing period for corn could be significantly shortened, this type of corn cultivation quickly pushed northward and made the demarcation line drawn by Young a thing of the past.

The rural landscape, dietary and agricultural traditions, and the entire life of southern Europe had been changed on many different levels thanks to the influence of corn. The traditional summer cereals—millet, sorghum, and panic grass—practically had been replaced by corn. Chestnuts, which before corn had been the dietary mainstay in times of scarcity or disaster, became a food of the past, a distant memory. Cyclical fluctuations of abundance and scarcity had disappeared with the advent of a cereal that ripened at the middle to the end of summer, a time of famine in the past. Perhaps because of that, and also due to both the failure on the part of naturalists to understand corn and the disdain for corn by the wealthy who preferred to eat wheat, corn undeservedly inherited a bad reputation as a lifeline in times of calamity. Corn came to bear the stigma of food for the poor. Effectively, corn was just that, both by virtue of its low price and its direct cultivation. Corn could be

Corn in Europe

the key to an ideal world with abundant food, an ideal yet to be realized. Poverty was rife in southern Europe and the gulf between classes overwhelming. Poverty, rural deprivation, and primitive living conditions all were intimately associated with corn.

To the north of the geographic limit for corn, the potato, another exceptional American plant, was the vehicle for a similar transformation. The potato, of Andean origin, thrived in the cold, damp northern European summers and was harvested in the fall. Although the potato carried less of a stigma than corn, it too bore the taint of a bad reputation. The first to adopt the potato as their essential staple were the Irish, considered by the rest of Europe to be a savage and primitive people. The introduction of the potato brought about changes just as profound as those associated with corn as far as diet, agriculture, and seasonal fluctuations in supply were concerned (Salaman, 1949). The history of the potato shared certain parallels with the history of corn. The potato thrived in a climate where mist was more the rule than sunshine. The societies that adopted the potato and corn for their respective staples were rife with inequality. Both plants provided the impetus for growth and, eventually, domination. The potato accompanied and provisioned new empires while corn was an accomplice in the decadence of older imperial powers, one more strike against it. Thus two American actors stood counterpoised. Those interlopers ironically became a central and intrinsic part of the history of Western Europe, the history of the center of the modern world.

A number of American influences contributed to a cluster of preconditions ushering in the period of revolution in Europe, an era of radical change. Among these figured potatoes, corn, and other American plants; New World plant products that were no longer considered exotic, such as sugar cane; precious metals; and the human merchandise of the triangular trade. Intellectual, political, industrial, and agricultural revolutions—all of them had been intertwined for more than a century.

The role of American plants in that process of accelerated change, frequently overlooked by historiography, figured prominently in two acute human tragedies. The first was the Irish potato famine. Potato blight ravaged the island's potato crops between 1845 and 1851. The widespread famine it provoked caused the death of one-fourth of Ireland's inhabitants and the emigration of as many again. The second, the appearance of pellagra in southern Europe, condemned hundreds of thousands of peasants to insanity and death over the course of the nineteenth century. Both gave an ominous glimpse into the dark side of progress.

CORN & SOCIETY BEFORE THE ERA
OF BOURGEOIS REVOLUTION

The adoption and dispersion of corn cultivation in Europe was a complex process. It had no simple explanation. There was no lone cultural agent responsible for promoting the introduction of the new plant. The migration of corn could not be attributed to a chance occurrence or some sort of fluke. Neither, on the contrary, could it be supposed to be part of an inevitable and natural process in which corn emanated outward in concentric circles at regular intervals. I did not find any reason to suppose that corn's diffusion in Europe was the product of a widespread and inexorable process, unless it was a process subject to more exceptions than rules. I came across coincidences and random incidents that could explain anecdotally how corn came to grow in a particular place. What I could not explain was how millions of people became involved in growing, trading, and eating corn. I searched, without success, for the ubiquitous and precise cultural agent to whom I could attribute corn's diffusion. Multiple collective agents, peoples and classes, appeared as important actors in the adoption of and resistance to corn. I was able to isolate certain natural and social factors that favored and promoted the introduction of corn, or limited corn's expansion, or even halted corn cultivation altogether. Among these, four stood out: growing conditions and the agricultural systems or their associated methods; population dynamics; trade, prices, and markets; and landownership and the relations of domination that pertained between landowners and direct producers.

None of these factors determined or subordinated the others. No direct hierarchy came to bear, although at certain times and under certain circumstances some factors were more compelling than others. Each factor was autonomous, with its own rate of change and transformation. In spite of this, they were not pure factors. Rather, they were tainted by their interaction with these and still other factors, factors that in this case were subtle, but that existed all the same. In the last analysis, such forces are abstractions, clusters of factors that behave predictably, that serve to better analyze and better understand a specific case or instance. They are provisional tools of knowledge that must not be construed as things or social subjects that explain much of anything. On this basis and with this caveat, I will try to trace how corn came to insert itself into Europe's natural and social landscape.

Strong direct sunlight, warm temperatures, and moisture were necessary prerequisites for growing corn. In Mediterranean Europe, the area where the American cereal first thrived, moisture and high temperatures were seasonally disassociated. Regions were characterized by dry summers with high temperatures and rainy winters with temperatures that, without being excessively harsh, were too cold for corn. Wheat and barley were the basic foods of the Mediterranean. Both were winter crops that were either dry farmed or sown in the fall, depended on seasonal fall and winter rains, and harvested in late spring. Lands in the Mediterranean were left fallow not only during the summer, but also during the winter after the crops were brought in. Crops were sown only every other year in order to replenish depleted soils.

Fallow land in the Mediterranean, just like everywhere else in Europe, was not the same as idle land. Uncultivated fallow lands still needed to be tilled and plowed in a several-step process, even though no plants were sown. These tasks served to eliminate weeds and preserve moisture, but especially to encourage nitrogen fixing and to protect soil fertility. A field that was left fallow in Europe was not the same thing as a field that was left fallow in America, where the land was neither sown nor plowed while it lay idle.

The continuous agricultural exploitation of the land and the traits of Mediterranean soils made the simple Egyptian plow, the *ard*, the minimal and fundamental farm implement in regional agriculture there. This lightweight plow was almost always pulled by a team of oxen. Parcels were grouped in fields that were sown at the same time and they had an almost square patterned configuration. Once the harvest had been gathered, the field remained open in order to serve as pasture for livestock. The raising of livestock, especially sheep and goats, was transhumant or itinerant and constituted a differentiated activity that at times worked against the best interests of agriculture. Nearly half of all arable soil was under cultivation in any given year, and as much again lay fallow. Perhaps a tenth of the available acreage was used for other purposes, such as vineyards, commercial endeavors, or complementary foodstuffs such as olive groves. In the densely populated Mediterranean there was no open natural agricultural boundary.

Since ancient times, wherever technically possible, irrigation works brought water to moisten parched Mediterranean soils. Irrigation made it possible to create areas of diversified, varied agriculture, intense in its use of both soil and labor. As a result of irrigation, citric fruits flourished. It was possible to introduce rice, tobacco, and sugar cane. Linseed, hemp, and even cotton were cultivated for textiles. Millet and panic grass were raised. The tremendous effort associated with irrigation works created privileged circumstances for growing corn. Warm temperatures and sufficient moisture coincided and under these optimal conditions corn was adopted in the region as a summer crop.

Corn varieties coming from the Antilles probably had short growing cycles. When these varieties were cultivated in the Old World, it was feasible for them to fit into the traditional crop rotation there, including fallow. Corn had definite advantages over millet and panic grasses in terms of yields and the length of its growing cycle. Those advantages ultimately led corn to eventually displace traditional Old World summer cereals altogether. Corn did not take the place of another new arrival—rice, also a summer grass—with which it coexisted. Even had corn and rice been competitive, rice required a larger workforce and a longer growing period, larger investment, and greater risk. Corn, with a lower demand for labor and very low investment in seed, produced more food per unit of cultivated surface area, making it much less costly and less risky. In the Mediterranean, corn did not occupy idle fields, but replaced and displaced other crops by virtue of its advantage over former irrigated crops.

The deep, rich soils of the great alluvial plain of the Po River valley received intermittent rains during the summer. This was unusual for Mediterranean climates, although it may have been typical of the region's culture and agricultural tradition. Since the High Middle Ages, a complex tapestry of drainage, irrigation works, and canals for barge transport had opened a domestic agricultural frontier for cultivation. Moisture and high temperatures were not disassociated there. The most advanced and complex agriculture in all of Europe was practiced in that region during the fifteenth century. The biannual alteration of winter grain and fallow was the basis for a thousand-year-old regional system of crop rotation. Corn's introduction in the sixteenth and seventeenth centuries changed land use and opened up the way for multicrop rotations. In multicropping, the cultivation of wheat or another winter cereal was immediately followed by the sowing of corn in the summer, which was harvested in time to allow the immediate planting of yet another, winter, crop. Corn cultivation served some of the same functions as fallow by preventing the proliferation of competitive weeds and impeding the loss of moisture typically brought on by tilling. The roots of the corn plant, much deeper than those of Old World cereals, did not drain away nutrients necessary for the proper development of winter crops. Any loss of nutrients caused by corn was compensated for by plowing under the corn plant's stalk and roots to act as fertilizer. In the short run, corn's introduction as green manure in multicrop rotations permitted the doubling of yields of traditional winter cereals upon eliminating fallow and added two summer harvests of the new American grain. In the long run, continuously alternating corn in the summer with wheat or barley in the winter led to soil exhaustion, which had to be compensated for with the growing use of fertilizer and the introduction of leguminous plants in new rotations (Sereni, 1997). There was no turning back from the elimination of fallow and the uninterrupted use of land, which worked in favor of territorial and land improvements.

The disassociation of high temperatures and moisture typical of Mediterranean climates was not so pronounced in Atlantic Europe. This region was subject to the influx of moisture from the ocean and the Gulf Stream that moderated winter and summer temperatures alike. There the lack of moisture attributable to irregular summer rains was not the crucial impediment to the cultivation of seed corn. Rather, it was the lack of warm temperatures that kept corn from ripening above forty-six degrees north. To the south of that boundary, agriculture and livestock, especially large livestock, were an integral part of peasant production. Agricultural tasks alternated between cultivated fields, where winter cereals tended to be sown, unirrigated fallows, and meadows where pasture grasses grew. Irrigation works probably were prior to and different from Roman or Arabic traditions. Before corn's introduction, where irrigation was necessary it tended to be for meadows rather than for cultivated fields. Such was the case in northeastern Portugal, where corn was introduced and spread in the sixteenth century. Corn was planted in irrigated, sloping meadows, arranged in terraces or platforms.

Later, in the seventeenth and eighteenth centuries, corn extended into unirrigated meadowlands of northwestern Portugal, northern Spain, and southwestern France. The transformation of meadows into cultivated fields of summer cereals raised productivity and increased the availability of foodstuffs, although it also probably upset the equilibrium between livestock and agriculture in favor of livestock. In those regions where the distinction between field and meadow did not refer to the type of soils or the source of moisture but only to the seasons—cultivated fields in winter and meadows in summer—corn's appearance upset that sequence and in many places opened it up to a multicrop rotation: wheat, corn, barley, or rye, corn, wheat, or to more intense rotation that preserved fallows, such as wheat in winter, corn in summer, fallow in winter, and corn in summer. In his travels through southern France in the 1780s, the English agronomist Arthur Young stated, "Where there is no maize, there are fallows; and where there are fallows, the people starve for want. For the inhabitants of a country to live upon that plant, which is the preparation for wheat, and at the same time to keep their cattle fat upon the leaves of it, is to possess a treasure" (cited in Crosby, 1972: 176).

In the climate of transalpine continental Europe, characterized by high levels of humidity and rain concentrated during the summer months, corn reached its northern limit. Corn was stopped short by brief summers that were insufficient for corn to ripen properly. To the south of that line, corn was established successfully. Longer summers and higher summer temperatures worked in corn's favor, even though moisture levels were not so high and rainfall could be irregular. The European corn belt was established to the north of the Balkans and in the Danube River basin, in southern and eastern continental Europe.

Before the second agricultural revolution of the eighteenth and nineteenth centuries, the three-field system had prevailed in transalpine Europe since medieval times. Fields were arranged in long rectangular shapes, known as the strip system, with a 1:10 ratio of width to length. The shape of the strips owed to the use of a heavy wheeled plow that turned the soil with difficulty even as it was drawn by teams of six or more oxen or very strong horses. Deep plowing enhanced drainage. This was in contrast to the more shallow plowing typical of the Mediterranean, which sought to preserve moisture by impeding capillary filtration. In the three-field system, one field was sown with winter wheat or rye for human consumption; in another, spring barley was sown for forage for draft animals and livestock during the cold winter months and for the elaboration of beer; the third remained fallow. Once the harvest was brought in, all three fields were opened for free pasturage of livestock, especially large livestock. There was no association between rights to cultivation and those of pasturage. Livestock belonging to anyone could pasture on any of the parcels. The village or community had an important role in the direction and supervision of agricultural tasks, which implied a high degree of cooperation among its members. The best-known historic model of the European peasant under feudalism was derived from such conditions in transalpine continental Europe.

Corn probably became part of the three-field system as a secondary spring crop. By virtue of its high yield, corn came to be substituted for barley as a forage crop in the south. Wine was substituted for beer as barley cultivation declined. Once corn became part of the traditional rotation, it displayed additional potential as a crop sown immediately after the wheat or rye harvest, just before the time a field lay fallow. Corn was tremendously flexible and it thrived in summer, a season for which there was no competitive crop in the three-field system. This allowed corn to occupy a new niche: as a third cereal crop after the winter cereals were brought in, or as a substitute for a spring crop. Such a practice considerably raised soil yields without eliminating fallow every third year. The introduction of corn only rarely brought about multicrop farming in southeastern continental Europe prior to the nineteenth century. Even in the late twentieth century, in many places corn became part of a field-fallow rotation (Fél and Hofer, 1969: chap. 2).

Corn's high yield was one of the most frequently cited reasons for its introduction throughout Europe. Conventional wisdom placed the average yield for corn prior to the nineteenth century at between two and three times the yield of traditional Old World cereals (Clout, 1977: 410). It was not easy to accurately assess European corn yields over extended periods of time. Before the nineteenth century, grain yields were measured as a ratio of units of seed utilized to units harvested. Data compiled by Traian Stoianovich (1953) for late seventeenth or early eighteenth century Romania placed corn yields in two tiers: one that ranged from twenty-four to sixty and another that ranged from forty to eighty units of harvested grain per unit of sown seed. The traditional European cereals yielded in a range of three to six per unit of seed. These yield figures for Old World cereals coincided with those compiled by B. H. Slicher van Bath (1963: appendix) even though, under some rare and especially optimal conditions, that range rose to ten per unit of seed.

The enormous difference in yields between corn and Old World cereals when measured as a ratio of seed to product made it likely that assessments which pegged corn yields at double or triple the yields of traditional European cereals referred to yield per unit area. There was not a great deal of information available on yield per unit area. A range of 0.22 to 0.67 tons of wheat per acre in eighteenth-century Europe would not have been implausible, although the majority of producers probably fell closer to the lower end of the spectrum than the higher end (Bowden et al., 1970: 58; Felloni, 1977: 1.2; Grigg, 1980: parts 2 and 3). Corn yields probably were anywhere between 0.45 and 1.3 tons per acre, although the higher-end yields were most likely an exception to the rule.

The difference between corn yields and yields for traditional European cereals was nonetheless considerable and without any doubt was one of the factors that worked in favor of corn being readily adopted. It should be taken into account that average European yields, measured as a relation between seed and product, did not rise between the sixteenth and the eighteenth century. In many cases these declined. It also probably was the case that the number of plants per unit area rose over the same period in order to compensate for any loss in yields. The introduction and expansion of corn cultivation took place during a time when yields for other agricultural crops in Europe remained stable. The high productivity of corn was all the more noteworthy in such a context.

Nevertheless, corn was not chiefly cultivated for its high yield. The main motive for adopting corn was the fact that corn constituted an additional crop. Corn utilized idle land and made it productive during the summer, previously a dead time for cereal production. The traditional European staple cereals—wheat, rye, barley, and oats—were winter or spring crops. Corn did not compete with them so much as complement them. Corn's high yield did not represent a marginal increase, but rather was over and above that of traditional cereals. This was not true of millet, panic grass, and sorghum, summer crops that were limited or secondary foods. In their case, marginal difference in yields would have accounted for corn replacing them. Additionally, corn's short growing period with respect to millet and panic grass made corn a noncompetitive, or better yet, complementary crop in the agricultural calendar for European cereal crops. There was no data available permitting comparison of corn's growing period with the growing period for millet and panic grass in Europe. In tropical Africa corn had an advantage of at least a month with respect to millet and panic grass, an interval that probably increased as corn cultivation moved further north. There was information on the presence and extension of early maturing corn varieties in Europe (Harasti da Buda, 1788). That data indicated that in all likelihood it was not only the difference in yields but also the short growing period that determined the substitution of corn for the lesser grains of summer.

In most cases, corn production represented an overall increase in soil productivity. This increase represented changes in the traditional systems of crop rotation. Land was worked uninterruptedly. In some cases, such intensive exploitation of the soil with the introduction of corn led to the total elimination of fallow and to multicrop farming. In general, corn in Europe was not a pioneer of virgin lands as it was in China. Rather, corn made a place for itself on seasonally idle lands that were under cultivation at other times of the year. Corn was a factor in the intensive use of scarce land. Southern European summers had been a dead time for food production. With the introduction of corn, summer became harvest time. Corn was one of the elements that shaped the agricultural revolution in Europe. That agricultural revolution was a prerequisite for other revolutions, indispensable in the founding of that liberal, industrial, urban, and modern world that became a model for progress.

A revolution in European agriculture took place between the sixteenth and eighteenth centuries. It was distinguished by a more intense agricultural exploitation of the land and the elimination or diminution of fallow. This characterization would not be acceptable to those who defined and defended the idea of a second agricultural revolution—the first would have been the invention of agriculture and its role in making populations more sedentary—as a technical process that raised yields per unit area, measured cyclically by crop. For those who supported this position, the second agricultural revolution was the result of the application of scientific knowledge to production. It was the diffusion of knowledge and the direct result of the effect that elites and intellectual vanguards had on society.

Such a definition of agricultural revolution is elitist and strictly agronomic. I propose, instead, the idea of revolution as a result of collective knowledge and collective action. This sort of revolution emerges from below. It consists of increases in aggregate production of a region measured in terms of prolonged cycles, rather than by annual individual crop yields. Revolution is conceived of as a social response, a collective response, in order to satisfy new necessities. An example of this might be changes in patterns and models for land use that enable a region as a whole to become more productive. The difference between the two definitions lies in the frame of reference in which the term revolution is placed. If that frame of reference is capitalist agribusiness, the agronomic vision of an agricultural revolution is legitimate. If, on the contrary, that framework is one of a society as a whole, then the agricultural revolution in Europe precedes the formation and application of scientific knowledge to agricultural production.

Around 1500, Europe had some 80 million inhabitants, a figure that was the same as in 1300. The virulence of the plague, the Black Death, cut short any tendency for demographic growth in the fourteenth and fifteenth centuries. Population growth over the previous thousand years had been slow but continuous. From 1347 to 1353, when the first epidemic reached catastrophic proportions, between a third and a fourth of Europe's population perished. Not until the sixteenth century was population growth once again widespread and by 1600 the total population of Europe was estimated at 100 million. The tendency for growth halted in the first half of the seventeenth century, only to begin once again after 1650, and reached approximately 120 million inhabitants by 1700. Population growth was sustained and accelerated in the eighteenth century, perhaps reaching 180 million people by 1800. Population growth exploded in the nineteenth and twentieth centuries, a process that has been dubbed "the vital revolution." Europe's population reached 390 million inhabitants in 1900 and 635 million in 1975 (McEvedy and Jones, 1978: part 1).

The long-term tendency for population growth since the sixteenth century did not follow the same pattern everywhere in rhythm and magnitude. Italy, for example, had the highest population density in all of Europe in 1300. It lost that distinction before the end of the nineteenth century to England, the Low Countries, Germany, and the northern countries of eastern Europe. The fastest population growth was recorded in eastern Europe. It was impossible to generalize about the pace of growth and the impact of European emigration. Fifty-one million people left Europe between 1850 and 1930 (Crosby, 1972: 216), and ultimately such a phenomenon came to accurately reflect a secular tendency.

Sixteenth- and seventeenth-century Europe was predominantly rural. Although it was difficult to pin down an exact proportion, the most commonly cited figures estimate that rural inhabitants accounted for at least four-fifths of the total population. In those circumstances, demographic growth led to greater pressure on the land. A process of mobilization and resettlement, of decreasing population in the mountains in favor of the lowlands and valleys, increased that pressure even more. The motives compelling such relocations were not clear, although climate change figured as one of them. There was no consensus as to whether or not a significant process of urbanization, like that which clearly took place in the nineteenth century, also occurred at this time, which would have had the effect of relieving some of the pressure on the land. Nevertheless, there was consensus that with few exceptions absolute agricultural population did not stop growing until the twentieth century. England was one of the places where the shift away from such a tendency took place somewhat earlier.

Few European countries had agricultural frontiers of virgin land during this period. Where such lands were available, agriculture expanded as these new lands were brought under cultivation. This especially was the case in eastern Europe and the Scandinavian countries located north of the upper geographical limits for corn. Nevertheless, it did not seem likely that in any country the agricultural frontier would have had the capacity to absorb all the new agricultural population. Neither were domestic frontiers sufficiently large to absorb the growth in rural population. In any case, their exploitation would have implied the reclamation of lands through regional transformations, which proceeded at only irregular intervals over the period. Agriculture was practiced in the context of shrinking per capita availability of land, as the demand for food and agricultural products rose due to demographic pressure and to changes in consumption patterns, especially of the better-off urban sectors (Dovring, 1965; Mauro, 1969: chap. 2; Mols, 1974; Helleiner, 1967).

We do not have a satisfactory explanation for the causes of that demographic growth. Conventional wisdom sees it as a result of modernization, a result of the application of scientific knowledge about health and living conditions or, once again, as a revolution carried forward by elites and vanguards. This interpretation is not warranted by the facts. The waning virulence and, later, the virtual disappearance of the plague preceded knowledge about its causes and treatment. The eradication of the plague owed partially to the formation and accumulation of immunities, and partially to gray rats replacing black rats, one of the carriers of the plague, as a dominant species. The incidence of pellagra in Europe lessened before its cause was widely known and even rose in the United States when there already was an effective treatment for it. Sanitary conditions in European cities probably never were worse than when the Industrial Revolution provoked massive urban growth in the nineteenth century. Until the end of the nineteenth century, there was no clear correlation between advances in scientific medicine and population growth. There still was not in most of the rural underdeveloped world by the late twentieth century.

Population growth was directly related to the production and availability of food, the old Malthusian dilemma. Until the late nineteenth century, the food supply for the growing European population basically depended on production on that same continent. But this self-sufficiency never was self-contained. Beginning in the sixteenth century sugar was produced overseas. Eventually this also was the case for other products increasingly important to European diets. With the exception of sugar, which would not become a maintenance food until the time of the Industrial Revolution, basic staples and especially cereals were produced on the same continent. This type of qualified self-sufficiency did not take place everywhere. Many regions or even nations had to systematically resort to imports from areas that produced a surplus. The traffic and trade in cereals was an age-old activity that spread and became more complex as a consequence of population growth and improved means of transportation.

It is likely that the demand for food grew at a rate exceeding the rate of population growth. This was a result of changes in diets, changes in food preferences, and changes in supply systems. Even though preindustrial Europe never was a paragon of widespread prosperity, at that time it already had the highest average level of consumption in the world, a reflection of its imperial power. In 1750, England had a per capita income that was three times that of Nigeria or India in 1960 and was the same as that of Mexico or Brazil in the early 1960s (Deane, 1965: chap. 11). Beginning in the sixteenth century, European diets improved and benefited from a wider array of foods, although they continued to be subject to enormous inequities and differences particular to distinct regions and classes.

For some authors, such as Ester Boserup (1965) and other neo-Malthusians, demographic growth was the independent variable in any relation between population and agricultural production. Demographic growth was the factor driving any changes in agriculture. Thus, population growth in agrarian societies was translated into technical changes in agriculture in order to more intensely exploit the land. In this scenario, the elimination of fallow was a response to greater demographic pressure on the land. For others, like William Langer (1963) or Alfred Crosby (1972), the ability to produce food was the independent variable. That is, any population growth was the result of a greater capacity to produce food. The introduction of edible New World plants, especially the potato and corn, increased European potential to produce foodstuffs and, hence, created a favorable atmosphere for a demographic revolution.

Two such mutually exclusive explanations are possible and suggestive when they are applied to nations or large social aggregates over prolonged periods of time (Grigg, 1980). It is likely that the interaction between population and food may be more complex on another level entirely. Peasant families in agrarian societies, the most basic unit of production and reproduction, provide new and different insights into the relationship between population and food. The concrete factors that regulate population growth for peasant families include their strategies for reproduction. These strategies take into account the availability of both land and labor; social and biological subsistence needs; relations of dominance that dictate the exaction or expropriation of a share of production; agricultural techniques and the botanical repertoire; together with decisions about family composition—celibacy, matrimony and the age of marriage, dowry and inheritance, residence and access to land. The study of reproductive strategies clarifies many things about population dynamics. As far as the social life of peasant families is concerned, apparently competing explanations can complement each other if they correspond to an issue in which both population growth and nourishment are integral components.

At the beginning of the nineteenth century, according to my own calculations, corn already had an impact on at least 40 percent of Europe's population. By that I mean that people both produced and ate corn, or ate other things that had been produced in association with corn within a crop system. Corn was commonly available to these people in the marketplace and as a dietary alternative. Corn was integral to the dietary system. This sort of influence of corn was not the same thing as dependence. The influence corn had did imply interdependence, inasmuch as corn was a part of the total dietary repertoire. The American cereal already had an important role in the reproduction of society, especially in its agrarian and peasant units, and corn was an important component in the equilibrium between population and subsistence. Corn's role loomed large as the nineteenth century progressed. Diets changed the most radically as population growth reached its most accelerated pace to date. Growing urbanization spurred the separation of producers and consumers. Until the eighteenth century, corn had been devoted primarily to direct human consumption in Europe. After that point, corn became instead one of the mainstays of meat-based diets.

Probably another similar proportion of the population found itself under the influence of the potato in the early nineteenth century. The cultivation and consumption of the potato had extended throughout northern Europe in a process that paralleled corn's, although subject to its own inner logic and dynamic. The potato, too, occupied land otherwise idle during fallow and changed the system of crop rotation in order to effect a more intensive use of the soil. European demographic growth relied on American plants. It relied on the participation of those plants in an agricultural revolution that intensified soil use by taking advantage of what otherwise was a dead time for the cultivation of crops. The vital revolution could be understood both as a prerequisite for and as one of the causal factors of the other transformations that shook Europe after the mid-eighteenth century.

A long-lasting inflationary tendency appeared early in the sixteenth century and took off after 1550. This was known as the price revolution. Grains and other agricultural commodities were the most affected by the movement in prices, which often tripled or quadrupled between 1500 and 1600. Real salaries fell over the same period, at times by as much as half. Those most affected by the drop in wages were probably the agricultural day laborers. They were good times for the agricultural producer, but much better for the owners of land and capital (Slicher van Bath, 1963). It was in that setting that the first experiments with corn and corn's earliest introduction to Europe along the Mediterranean rim and the Atlantic coast took place. The new crop not only had certain agricultural advantages but also stepped in and closed a gap in the food supply as a relatively cheap maintenance food. Corn became a food for the poor, responding to the demand of Europe's largest and least splendid market. The very stigma of poverty provided the conditions for the expansion of the consumption of this American grain.

There has been much discussion about the causal factors behind Europe's prolonged period of inflation. For many authors it signaled demographic recovery that stimulated an increase in demand. For others, the arrival of precious metals from the New World brought about a revolution of prices with the sudden expansion of the monetary base and the circulation of money in Europe. In this case, two apparently mutually exclusive hypotheses could be complementary. In any case, America's sudden appearance on the European scene, in the form of precious metals as well as plants, remains a factor in any explanation.

The sixteenth-century inflationary tendencies began to slow around 1600 and the prices of wheat and other cereals remained more or less stagnant until 1640. In the second half of the seventeenth century the prices of maintenance foods fell. Some years these were fully half what they had been earlier in the century, although never approaching levels predating the revolution of prices. Prices remained relatively stable in the first half of the eighteenth century, at levels that were at least double those of the fifteenth century. An inflationary tendency reappeared in the second half of the eighteenth century and lasted a hundred years. Over that period of time, the prices of wheat and maintenance foods reached and surpassed the highs that had been recorded in the early seventeenth century (Braudel and Spooner, 1967; Slicher van Bath, 1963).

The behavior of long-term prices showed great variations over time and space. Sixteenth-century prices for wheat and other cereals were appreciably higher along the Mediterranean rim. This may partially explain the incentive to introduce and experiment with corn in that region as opposed to western or eastern Europe. The trend reversed itself during the eighteenth century and western Europe, having recently come into a good deal of money, paid more for its maintenance foods than countries of the Mediterranean rim or eastern Europe, which continued to enjoy lower prices.

In spite of any variations over time and space, the long-term upward movement of prices was a generally valid tendency. This trend illustrated another important phenomenon: the expansion of a monetarized mercantile economy between the sixteenth and eighteenth centuries. This expansion came about, to an important degree, at the expense of subsistence economies. Peasant economies always preserved a degree of self-sufficiency. By the same token, their dependence on the market and money rose. The bond between peasants and the market was universal. However, very different levels of intensity and different mechanisms of articulation pertained from peasant econ-

Corn & Society

omy to peasant economy. Almost all these were associated with forms of dominance, with mechanisms of control over land and capital. In the nineteenth century many European peasants still depended on money very little. They were capable of satisfying almost all their needs using their own devices, even though they might have to surrender nearly half their production to landowners and, indirectly through them, to the market in payment for ground rent.

The European expansion that began in the second half of the fifteenth century affected European markets in other ways as well. It modified trade routes and changed points of embarkation. The Mediterranean, which had been the axis of commercial traffic since ancient times, was displaced by the Atlantic and the North Sea. New Atlantic routes linked America, Africa, and the Far East. North Sea ports were home to new and more dynamic financial, commercial, and manufacturing centers. The Mediterranean had been fragmented by the expansion of the Ottoman Empire and was completely marginalized by the new transatlantic routes. As a result, the Mediterranean lost access to a great part of the surplus that it had formerly handled in its role as commercial intermediary and through the exportation of manufactured goods, especially textiles. Suddenly the Mediterranean world had to cope with its food problem by falling back on its own agricultural development. A century-old trend that extracted resources from the countryside for their reproduction in financial and commercial activities had to be reversed. During the course of the sixteenth century, commercial and urban capital was invested in agricultural production, especially in north central Italy. The land always had been a source of wealth, but now agriculture became a business, a means for the reproduction of capital. Corn's short growing cycle, its high yield, and the utilization of idle land that corn cultivation proffered were readily exploited within the logic of the reproduction of capital.

Economic depression and slowed population growth affected all of seventeenth-century Europe. The effects were even more harshly felt in Mediterranean countries. While economic and population growth rebounded in the eighteenth century, the response to this was not the same everywhere in Europe. Southern Europe declined or at the very least was marginalized from the forces that prompted the Industrial Revolution in western Europe. Conditions were being created for the accelerated reproduction of capital in western Europe, and capital from the world over was drawn to the imperial metropolis. Southern Europe began to be considered backward and their agricultural practices deemed primitive. Even though there was not much evidence to support such a conclusion, the idea persists to this day.

Between the sixteenth and eighteenth centuries, vestiges of feudal Europe profoundly influenced landownership and access to land. The nobility, a hereditary class, controlled access to land. In exchange for the use of the land, the nobility received divers payments in the form of money, work, stipulated services, or a share of agricultural production. The church, especially the Catholic Church, was a very important landholder, with a status similar to that of the nobility, and in some cases also wielded the powers of state. In addition, various European churches enjoyed the right of appropriation over a share of total agricultural production through tithing. Incipient nation states had direct control over land belonging to the Crown and levied taxes on the production of territories under their political sphere of influence. There frequently were tensions and contradictions between the forces that, under many different guises, disputed the appropriation of rents and peasant surpluses. These forces were protected by hereditary rights and also accrued other, new rights as well. These included operating as intermediaries in commodity markets or in the exchange of money. These forces formed a network to forcibly extract surplus from peasant producers. Under a wide array of circumstances, these peasants were compelled to transfer significant proportions of their production or their labor in exchange for occupying and working the land. All the available figures placed the amount for rents levied on peasant production or work at half of total peasant earnings, at least.

The appropriation of peasant surplus and peasant labor was varied, complex, and carried out through the intervention of multiple intermediaries. A study made by the Napoleonic forces of occupation in Sicily, for example, documented 1,395 feudal rights and privileges that involved exclusions and prohibitions applying to peasants or payments by peasants to overlords (Bowden et al., 1970: 166). Many of the complex feudal norms had lost their practical importance or their economic purpose and were no longer practiced in eighteenth-century Europe. Others were strengthened or new norms were created that favored the overlords. Feudal control of the land, exercised in various forms, did not diminish in importance in most of Europe until the nineteenth century, even though new forms of appropriation had already arisen, including forms that challenged the privileges of the nobility and ecclesiastic institutions. The original medieval definition and sense of feudalism underwent profound changes everywhere. Feudal rights became instruments for mercantile relations in the circulation and accumulation of capital by several means. Feudal rule, directly or indirectly, was transformed into relations of landed property, in the liberal meaning of the term. The overlord or noble, who in the medieval era was fundamentally a public authority, became a private landowner. Thus, control over land by the seigniors was severely compromised and suffered a loss of hegemony in many regions during the sixteenth and eighteenth centuries. Despite this, seigniorial land rights remained determinant in European agrarian relations.

It is not possible to generalize about the enormous diversity in feudal rule and its transformation. Typically, two models have been utilized in order to describe and analyze the transition from noble to landowner. There was either the productive, direct administration of the land by the overlord or indirect administration, in which production remained in the hands of sharecroppers or tenants and the overlord received rents in money or goods. The line dividing each prevailing model was set at the Elbe River. To the east of the Elbe, export agriculture was practiced. There, direct administration, also known as the Junker system, predominated. The overlord directly worked the land for his own benefit and any exactions from the local peasantry was in the form of prestation of services or the provision of labor power under servile conditions. To the west of the Elbe the indirect administration of land was the rule. Sharecroppers, croppers, or tenants, ranging from lowly peasants to rich mercantile producers, paid in money or in kind for the use of the land. The dual model is useful only if it is taken into account that predominance of one form or another did not imply exclusivity and that in practice such abstract characterization rarely applied in a pure form (Maddalena, 1974: 287–304).

Sharecropping or leasing were quite common on the large estates to the east of the Elbe, although that privilege was reserved for well-off or rich peasants who owned a plow and a team of draught animals with a minimum of six oxen or horses. Ownership of a plow or of draught equipment was associated with a class of rich peasants that did not have to provide personal services. Typically, personal services were in payment to the overlord for the loan of the draught equipment. Rich peasants also could pay rural workers to provide personal services to the overlord on their behalf. Poor peasants, who provided stipulated services for the operation of large estates and manor houses, and who received only a paltry payment in money and in kind, also had the use of a small plot of land to work for their own benefit. The distinction between peasant and agricultural proletariat or rural peon was not as clear as it would become later.

To the west of the Elbe, in many cases, overlords directly administered a part of the territory over which they exercised control, but for the most part handed over the rest to sharecroppers. The provision of stipulated services had not disappeared entirely, although their importance had lessened with the advent of rents in money or in kind. In some cases, nobles were absentee landlords in every sense and handed over all the land to an intermediary who divided it and sublet it to local peasants. In many areas, the tenants or share-croppers had acquired a right to remain and even inherit the land through emphyteusis, a practice sometimes indistinguishable from ownership. On the other hand, the medieval institution of open fields once the crops were harvested was preserved in most of western Europe until the nineteenth century, especially where livestock yielded better profits than agriculture. Powerful noble corporations, like the Mesta in Spain, constituted a formidable obstacle to all attempts to fence the land.

The enormous variety in the forms of administration and territorial control did not disrupt the long-term tendency toward transforming feudal rule, a political relation, into the private appropriation of land, a juridical-economic relation. It was in England where that tendency advanced most rapidly. Over-lords became absolute landowners, to the detriment of both the rights of those occupying the soil and of the tradition of common fields. The enclosure movement attempted to administer land directly. Livestock also was confined within the fenced spaces and effectively became part of the agricultural production process. At the end of the eighteenth century, fenced land already constituted the overwhelming share of rural property. As that tendency continued to progress, servile relations for the provision of personal services disappeared and salaried relations developed until in the end an agricultural proletariat emerged as the leading rural class. Seigniorial rights were transformed, in that case, into access to property for the benefit of nobles and rural overlords.

The French Revolution abolished feudalism in favor of private appropriation in late eighteenth century France. Those who benefited most from the new form of ownership were not nobles, but tenants or sharecroppers. Many of these tenants and sharecroppers were prosperous mercantile producers, the rural bourgeoisie of the Third Estate, and well-off peasants. Elsewhere in Europe, the functional transformation of feudalism into capitalism came about without a clean break with the past and without direct expropriation until the agrarian reforms of the late nineteenth and early twentieth centuries. Common fields lasted until that time, together with specialized and trans-humant stockraising. In many cases, attempts by the Napoleonic forces of occupation to impose French land reforms elsewhere in Europe ultimately backfired. In Italy and Spain, nobles and former large landowners participated in national resistance movements against the reforms imposed by the French. Once the invading French forces had retreated, anything associated with the French, such as land reform, was seen in an unfavorable light.

From the point of view of the majority of poor peasants, this long and complex process ultimately ended in dispossession or even more intense exploitation. Land rents demonstrated a clear tendency to rise and conditions promoting stability and long-term tenure diminished. Lease agreements were for shorter periods of time and their conditions became more onerous. In some cases the surface areas available for sharecropping or cropping also decreased, and rural workers were denied access to even small plots. The division of land into smaller parcels in many cases implied inadequate production. Peasants became unable to keep back enough of the harvest to satisfy their basic consumption needs. They were forced to make up for it with the sale of their labor power. Salaries of rural workers fell as part of a long-term tendency. In eastern Europe the conditions for the provision of servile labor became more harsh. The intensification of land use and of labor became the only way the rural poor could survive. In spite of the intensification of agriculture, European peasants suffered a marked decline in their standard of living and nutrition. Vagrancy and marginality of peasants forced off the land became widespread and were met with repression. European social structure became more complex and distances between different classes and groups increased. Disparity became increasingly pronounced. The rural population, never more numerous, lived at the height of the Age of Enlightenment in conditions that were worse than those of the Dark Ages.

The diverse combination of long-term forces and tendencies led to diverse settings for the introduction and propagation of corn cultivation. Different groups took the initiative in introducing corn as a function of their own interests. So, in the Balkan countries and the Danube River basin, it seemed that landowners took the initiative in adopting corn. Turks in the south and Christians in the north directly operated their large estates and sold their agricultural product on the export market. Corn was not initially part of the export market. Corn was utilized to attend to the needs of those same large estates, as feed for livestock and nourishment for the servile workforce, who planted corn in idle fields in the off-season. Corn freed up a greater proportion of the more expensive cereals for the export market. Thus, profits for the owners of the large estates grew more than is represented simply by the price of corn alone. The peasants could not successfully resist the introduction of the new grain. Corn deprived them of their traditional foods and increased their workload during what had been less labor intensive periods of the agricultural calendar. Any resistance was met with force and even repression. Corn was aggressively promoted in this case by the overlords because it directly served their interests.

It would seem that something similar happened in the surrounding lands and the Lombard plain. Harasti da Buda (1788: 12) testified that peasants showed an indescribable aversion to adopting corn, which they feared would do harm to the land. In the complex agrarian structure of those areas, corn not only was introduced in lands directly administered by overlords or by mercantile farmers, but corn also was imposed on the croppers and peasant sharecroppers in order to supply regional markets. Peasant resistance to the new crop was found in the context of its imposition. The prices of wheat and of other cereals rose in that region more than in the rest of Europe during the very years corn made its appearance. This was at the same time that the capacity to export manufactured goods and the importance of the Mediterranean as a center for trade waned. The need to decrease the volume of expensive imported cereals created a niche for the cheaper substitute—corn. Mercantile producers or landowners imposed corn cultivation. It was they who then sold peasant production paid as ground rent on the open market. Their actions received state backing. Nations understood that self-sufficiency in the production of cereals was a national security issue, a reason of state.

In contrast, the introduction of corn in the Atlantic region seemed to owe fundamentally to the initiative of peasant producers in an economy in which agricultural production was integrated with the raising of livestock. This was the case in the Balkans and north central Italy. In the Atlantic region, the scarcity of cereals for part of the year opened a niche for corn. In terms of land use, corn did not compete with other less productive cereals but, rather, with meadows used only for pasturing livestock. Corn stalks and leaves used as forage essentially compensated for any loss of pastureland, and the grain itself, which fed people and eventually livestock in periods of crisis, yielded a net profit. Under those circumstances, the peasants became enthusiastic propagators of the crop.

Whether by imposition or by their own initiative, corn cultivation soon became a part of peasant tradition. It formed part of their technical and botanical repertoire. As a part of this repertoire, corn demonstrated new qualities, new uses, and potentials. Frequently, corn's introduction meant that larger proportions of more expensive commodities were freed up for the market in order to meet the rising cost of rents, inasmuch as the new grain contributed to the self-sufficiency of the productive unit. The potential rose for creating mercantile surplus or sending greater proportions of traditional crops to market. On the other hand, production for self-sufficiency was shifted to summer, the season that before only marginally occupied the peasant workforce. It led to the intensification of labor by more evenly distributing agricultural tasks over the course of the year. In a contradictory way, it favored a more intense integration with the market, at the same time that it reorganized and reinforced self-sufficiency.

The new grain ultimately became part of the peasant diet in those regions where it was cultivated, eventually becoming its most important component. Stone-ground flour cooked in water, most of the time with nothing more than salt, became the everyday food, the daily bread. This corn paste was seasoned with milk, oil, or butter where and when those were available, with other ingredients tossed in on special occasions and religious holidays. Corn prepared this way was known as polenta in Italy, as *mamaliaga* in the Balkans, and as *puchas* or *migas* in the Iberian world. With many names and with a thousand combinations, corn paste became the most common and at times the exclusive food of poor peasants. In cities, corn was mixed with other bread flours or was consumed increasingly like paste, especially by those with no other recourse. Corn became the bread of southern Europe's poor.

There was a close association between corn and poverty on this rural continent. As might be expected, however, there was another, third party to this association: wealth. Corn generated wealth for landowners, shopkeepers and moneylenders, overlords, and the new middle class. Ironically, most of those who stood to gain the most from corn never even ate it. They ate wheat instead. Corn was a commodity with an unlimited market: that of the poor, who made up the vast majority on a continent that was bracing itself for the coming of modern capitalism.

: 10 :

THE CURSE OF CORN IN EUROPE

Corn was the chief food and perhaps the foremost agricultural crop in the principality of Asturias in eighteenth-century Spain. Its introduction was attributed to Gonzalo Méndez de Cancio, governor of Florida. The corregidor of Oviedo grumbled in 1769 that "this province produces no more grain than it needs, whether due to the poverty of the farm workers, who are all or mostly come from isolated farms and holdings belonging to others and work only those lands necessary to earn the yearly payment of their rent and to maintain their families with the bare necessities" (Gómez Tabanera, 1973: 178). In 1757 representatives described the principality's poor subsistence economy to the king in this way: "They do not have money (something rarely seen in the principality), but only that product of their own meager harvest or industry, that they barter or give in exchange for what they need for their own sustenance, reduced not to splendor, but to satisfying needs only just enough to survive" (Gómez Tabanera, 1973: 179).

About 1730 Dr. Gaspar Casal practiced medicine in Asturias. He had been appointed to reside in Oviedo, the only urban center of any importance in the entire province, which had only 1,748 regular inhabitants and 623 members of the clergy in 1757. There he came across a disease he was not familiar with, which the inhabitants of the region called *mal de la rosa*. Dr. Casal described the symptoms of that disease with skill and accuracy. Subsequently, these symptoms would be summed up as the three Ds: dermatitis, diarrhea, and dementia. Dermatitis, the most characteristic and visible of the symptoms, appeared on the neck and any extremities exposed to the sun, forming a hard red crust on the skin there. The skin irritation, for which the disease was named, appeared in the spring and faded during the summer and winter months. Digestive tract disturbances were seen constantly in the same patients affected by skin irritations. The mental affliction manifested itself in the early stages of the ailment as loss of appetite and listlessness, and in the terminal stages as insanity. One of the advanced signs of the malady was uncontrollable vertigo, known as St. Vitus's Dance. There was no cure for the disease and once the advanced stages were reached the condition was almost always fatal.

Dr. Casal had never seen anything like it. Perplexed, Casal was cautious with respect to forming any opinion as to either its causes or cures. He looked for clues in the diet and found that corn was the primary staple food of those who were stricken with the ailment. Those who had the disease consumed very little fresh meat, although their diet was complemented with eggs, chestnuts, nuts, dairy products, pears, and apples. He also discovered that while the peasant diet was virtually the same all over the province, *mal de la rosa* was not found everywhere in the region. Rather, the disease was concentrated in four localities. On top of this, Casal pointed out that his old friend and patient, Juan Bautista Dolado, never ate meat or fish but only bread, cheese, legumes, and fruits, which did not hinder him from reaching the ripe old age of eighty in fine health and without ever having contracted the dreaded *mal de la rosa*. The idea that the origins of the malady were located in the diet was, therefore, not conclusive. Nevertheless, Dr. Casal made note of a detail that remained undisputed: those afflicted by the disease were all poor peasants (Major, 1944).

Dr. Casal was the first to describe this malady, which later would become known as pellagra and would batter southern Europe in the nineteenth century. Nevertheless, he was not the first to publish something about the new affliction. François Thiéry published an account of *mal de la rosa* in Asturias in a French medical journal in 1755. In that article, the French doctor readily acknowledged that he relied on information he had taken from Casal. Casal's work was published posthumously in 1762 with the title *Historia natural y médica de el principado de Asturias*. That work included an illustration of a classic Adonis-like figure, flawless and in glowing health, marred by a conspicuous rash typical of *mal de la rosa* on the neck and forearms, the stigma of the pellagrin.

Dr. Francisco Frapolli published a brief monograph on pellagra in Milan in 1771, without having read either Casal or Thiéry. The peasants of Lombardy knew the same disease that the Asturians called *mal de la rosa* as pellagra. Frapolli agreed with Casal that those affected were poor peasants who ate almost nothing more than polenta made from corn and other grains. He came down more firmly but less unabashedly than Casal in favor of ascribing the malady to poor diet in which corn played a predominant role. Frapolli was the first to publish the name "pellagra," by which the disease would go down in history. Quite correctly, but perhaps relying more on intuition than on scientific rigor, Frapolli added that they were dealing with an entirely new disease, unknown in the past (Major, 1945).

Dr. Giovanni Battista Marzari established a clear correlation between pellagra and a diet composed almost exclusively of corn in 1810. Marzari stated that the malady was caused by the lack of foods of animal origin and, specifically, by corn. He added that pellagra was not hereditary or contagious. This directly contradicted earlier speculations by Professor Buniva, who conducted experiments in 1808 that involved inoculating himself with the saliva of infected patients. Marzari was considered to be the founder of the Zeist school, an intellectual current that attributed the cause of pellagra to the consumption of corn (Carpenter, 1981; Roe, 1973: 55–68).

Corn did not have a good reputation as a food at that time. Rather than dissipating, the old prejudices disseminated by sixteenth-century naturalists and doctors had multiplied. The German poet and essayist Goethe did nothing to dissuade corn's detractors with his vivid descriptions and keen observations on his trip to Italy in 1786. The sickly appearance of the Italian peasants living just beyond the Brenner Pass, which divided the German Tirol from the Italian Tirol, troubled Goethe. He attributed their pathetic condition to their diet that was composed almost exclusively of polenta: a thick mush made from flour cooked in water. Yellow polenta was made with corn flour and black polenta with Saracen flour. The Italians ate polenta plain, perhaps sprinkled with a little grated cheese. The German peasants on just the other side of the pass also ate polenta, but many other foods besides. They separated the corn mush into small pieces that they fried in butter. The Italians ate little else than unadorned polenta: meat at year's end, pears and apples in season, and green beans boiled in water and dressed with garlic and olive oil. When Goethe asked the innkeeper's daughter what the Italian peasants did with their money, she replied that they had masters who concerned themselves with that. Goethe added a personal touch to these observations when he confirmed that corn polenta caused severe constipation (Roe, 1973: 39–40). Certainly corn's reputation as any sort of delicacy suffered greatly from such remarks by that influential writer.

In the late eighteenth century, when corn's reputation as food plummeted, its prestige as an agricultural crop skyrocketed. Growers with practical experience cultivating corn and enlightened agronomists, usually nobles, who wrote about the plant both touted corn's virtues: its high yield and low cost, its growth in the summer or whenever the cultivation of traditional cereals was out of season, its use as green manure in eliminating fallow with no corresponding reduction in the yields of traditional grains, and, last, the impact of the more intensive use of the soil, of capital, and of labor. Some agronomists emphasized corn's role as food in their monographs, its abundance, and its low cost, although not necessarily its flavor or its nutritional value. With that campaign and under pressure from social and economic forces, corn cultivation spread rapidly. Corn displaced other traditional summer crops and was responsible for the reordering of cereal rotations. It increased profits for landowners and heightened the exploitation of the peasants who cultivated corn, the very people who increasingly depended on it for food. The cultivation and the abundance of the American cereal changed their diets. Corn's loss of prestige, promoted either by those who did not eat it or by those who studied the causes and effects of pellagra, was not serious enough to counteract corn's growing importance as food for the poor. Corn's bad reputation stubbornly persisted and served to reinforce the association between corn and poverty. The stigma of poverty indelibly branded corn.

The Zeist school, which blamed pellagra on corn, became fashionable in the nineteenth century, backed up by circumstantial data and facile correlations. Two currents emerged over the course of the century. The first, which prevailed in the first half of the century, considered corn to be an incomplete food, lacking in the essential building blocks of good nutrition. This current attributed pellagra to inadequate diets in which corn was the main component.

The work of Filippo Lussana and Carlo Frua, *Su la pellagra*, published in Milan in 1856, illustrated this current within the Zeist school. The authors held that corn was a food lacking in proteins, since its protein content was only 12 percent. Today we know that even that figure was exaggerated by 2 or 3 percent. In any case, at that time it was believed that proteins were essential for repairing the wear and tear on internal organs and should constitute a fourth or at least a fifth of total nutrition. Corn's protein content was well below that threshold. Lifestyle changes associated with modernization especially had affected peasants. They were subject to a more intense work pace that raised their overall protein requirements. At the same time, peasant diets changed in order to survive almost exclusively on corn consumption. Peasants had become victims of the new malady for that reason. The authors observed that crises and famines always had existed, but not pellagra. For this reason they concluded that the epidemic stemmed from historic nutritional crises resulting from a lack of calories as a source of energy, inasmuch as the new disease owed to protein deficiency. Their explanation sounded very close to what is believed today, albeit an unsophisticated and overly broad version of it, but it was not its equivalent. It revealed the general causes, but did not suggest the specific causes. The authors conceded this when they wrote at the beginning of their work that the lack of understanding of the physiological impact of nutrition had brought about an inevitable clash between direct observation and its amicable interpretation. The clash would persist for more than half a century and in some respects continues until today.

The other current, influenced by positivism, attributed the cause of pellagra to a specific substance, a toxin, produced by the decomposition of the corn kernel. The most outstanding proponents of this theory were Théophile Roussel and, especially, Cesare Lombroso. Roussel was a French doctor who did research on pellagra. Lombroso was widely known for his work on the relationship between criminal personalities and physical traits, a dangerous line of thought that still had adherents and that left prejudicial aftereffects in its wake. Lombroso devoted twenty years of his life to the study of pellagra. This current dominated discourse during the second half of the nineteenth century, when the disease reached its maximum incidence and its greatest resonance, as much in scientific thought as in debate over public policy aimed at combating it.

The school of thought that attributed pellagra to spoiled corn and the toxins that formed as corn decomposed owed much to the work of Ismael Salas, a Mexican physician. Salas received his graduate education in Paris and wrote his thesis on pellagra, which he defended in 1863. In it, he began by recognizing that the disease only appeared where corn was the principal component of the diet, but added that in many other instances in which that was the case pellagra did not crop up. The correlation between corn and pellagra was positive, but not generalized. Salas illustrated his argument with the case of Mexico, where the poor lived almost exclusively on corn without ever being afflicted with pellagra. For this reason, he concluded that neither corn nor its low protein content could be responsible for the disease. In order to support his conclusion, he compared the content of nitrogenous materials in corn, as proteins were identified in those days, with that of other grains. In that comparison, corn was equivalent to wheat, rye, and barley and easily surpassed rice. Nevertheless, pellagra did not appear among peoples with rice-based diets or diets dependent on the most common European cereals. Those contradictions suggested that there was a specific causal agent of pellagra. Salas tended to favor *verdet*, a term that referred to the odor of spoiled corn and by extension to the discoloration and stains that the grain showed in a decomposed state.

Doctor Salas confirmed that where the spoiled grain was not eaten, pellagra did not turn up. Additionally, Salas credited the method for preparing corn as a preventative measure to fight the disease. He pointed out that pellagra appeared where corn was ground raw and dry. Where corn was cooked in water saturated with an alkali such as lime or rock salt, the disease did not appear. Once again this sounded very close to modern assessments, but the two explanations are not equivalent. Salas maintained that alkaline preparations eliminated or neutralized any toxins released as corn decomposed. Polentas or dishes resembling polenta were not prepared this way, so corn toxins were released into the food as corn spoiled. Today we know that pellagra has nothing to do with any purported toxins produced by decomposed corn. Salas defended corn consistent with his own experience with corn in Mexico. Minimally, Salas exposed weak points in competing explanations maintaining that corn was a food lacking in proteins.

The debate between currents within the Zeist school and between the Zeists and their few detractors was not resolved in the nineteenth century. On the contrary, the debate became increasingly heated and dogmatic, even ideological, and entrenched around old arguments. Despite the ardor of the debate, it did not promote a critical and vigorous experimental current. Researchers were obligated to affiliate themselves with one position or another, to fall in line and to bow to pressure, to chose a side and run with it. In retrospect, many of the hypotheses and arguments had a great deal of depth and even good instincts in terms of what we know to be true today, but lacked rigorous experimental standards and continuity. Excesses were committed. Lombroso definitively identified the causal agent of the disease as pellagrosein. His conclusion prevailed, even though his experimental results could not withstand any degree of rigorous scientific scrutiny. The Zeists did not discover any effective treatment for pellagra, and the disease continued to follow its natural course.

Documented medical reporting confirmed that the geographic distribution of pellagra in the first decade of the twentieth century was far flung. The true extent of the disease was possibly wider still, but there were no medical reports or accurate diagnoses to back up such a conclusion. Doctors Claude H. Lavinder and James Babcock, the English translators of the book by Armand Marie on pellagra, identified three categories of the incidence of pellagra: endemic and cases relatively few, endemic and cases relatively numerous, and sporadic (1910: 75–76). The distinction between the first two categories could be determined utilizing figures and studies subsequent to the publication of the account. The category of few cases could be set roughly at around 2 or 3 percent of the total population. The category of numerous cases affected at least that percentage and in some districts more than 10 percent of inhabitants in those years in which the disease reached its maximum level of virulence.

Under the category of endemic presence and numerous cases were north central Italy, especially Lombardy, Venice, and Emilia; Walachia and Moldava in Romania; the Austrian Tirol, the same region where Goethe considered the inhabitants to be so prosperous and well fed; the island of Corfu in Greece; lower Egypt and the area bordering the Red Sea; and the provinces of Asturias, Lower Aragon, Guadalajara, and Burgos in Spain, although the lack of information on Spanish cases was the subject of angry debate; and southeastern France, including Gironde, the Landes, Hautes-Pyrénées, and Pyrénées-Orientales. The incidence of pellagra in Upper Garonne, that recorded numerous cases in the mid-nineteenth century, had fallen by the beginning of the twentieth century and was reclassified as belonging to the category of few cases. This was the first documented instance of pellagra coming under control.

The regions in which pellagra had an endemic presence with few cases included northern Portugal, the eastern portion of the Austrian-Hungarian Empire; Croatia, Dalmatia, Bosnia, Transylvania, Herzegovena, Bucovina, and Galitzia; Serbia, Bulgaria, and southwestern Russia, including Poland; Greece, Turkey, southern Italy and its islands, Algeria, and Tunisia. Only exceptionally did corn figure prominently in local diets, and the ailment did not turn up or simply was not documented. Otherwise, the geographic distribution and the intensity of pellagra at the beginning of the twentieth century was a direct reflection of the geographical location and importance of corn cultivation in Europe and the Mediterranean rim.

Pellagra was limited to Europe and the Mediterranean rim in the nineteenth century. By the early twentieth century, pellagra had struck the United States. Pellagra also had cropped up among mine workers in South Africa by that time. It is still debated whether or not the disease that struck in the English Antilles was really pellagra or not. Those primarily afflicted with the ailment were workers on sugar cane plantations whose diet consisted mostly of corn. Pellagra also had shown up in Mexico's Yucatan peninsula in the 1880s, years after Doctor Salas defended his thesis. Doctor Geoffrey Gaumer, of Izamal, attributed the appearance of pellagra to the consumption of imported corn between 1882 and 1891. Recourse to imported corn was part of the aftermath of a severe locust infestation that destroyed the corn crop in 1882. According to Gaumer, imported corn transported in the holds of vessels as ballast had badly decomposed by the time it arrived in Yucatan and its consumption caused pellagra. Between 1892 and 1901, Yucatan produced sufficient corn for home consumption and new cases of the disease did not develop. Between 1901 and 1907 the corn harvests failed again or were insuffi-

138   *The Curse of Corn*

cient for local needs, and corn had to be imported once more. The corn arrived in a deteriorated state and pellagra was the result. The malady came to affect up to 10 percent of the population in 1907, according to Gaumer. Gaumer mentioned that corn imports in the first decade of the twentieth century owed not so much to natural disasters as to the fact that the planting of hemp made for more attractive earnings for *hacendados* than corn (Marie, 1910: 69–70). Doctor Gaumer failed to mention the devastating effect that the hemp monoculture had on the diet of bonded and indebted peons. These individuals had to buy all their food at company stores at exorbitant prices and did not have access to complementary foods typical of the complex system of mixed farming in traditional corn fields.

The natural course of pellagra, its effective social history, varied in the different regions in which the disease reached its highest incidence. Pellagra was first documented in southwestern France in 1829. Until the mid-nineteenth century, the number of cases tended to rise. At that point, the government embarked on an energetic public works campaign that included the construction of drainage works and the reforestation of marshy areas. This had the dual effect of creating jobs and reclaiming new lands for agriculture. At the same time the division of large land holdings took place and small peasant property holders arose or were strengthened under the aegis of revolutionary legislation. Agricultural production was restructured and diversified. This raised incomes and improved the distribution of wealth. Cultivation and consumption of corn were not stopped short, but corn's relative importance, which had been overwhelming, began to diminish. The food in peasant diets became more varied and plentiful. The incidence of pellagra fell in the third quarter of the nineteenth century. No new cases were turning up by the last quarter of the century, although many of those already afflicted suffered the last stages of the disease. At the beginning of the twentieth century, endemic pellagra was essentially eradicated from southeastern France. Doctor Théophile Roussel, one of the most ardent defenders of the theory claiming decomposed corn caused pellagra, was credited with an important role in the eradication of that disease in France. The part he played was certainly significant, but perhaps ultimately he had been less effective as a clinical researcher than he was as a catalyst for popular outcry and public intervention in order to promote improved living conditions in the region (Marie, 1910: 340–49; Roe, 1973: 46–54).

The history of the ailment was unclear in Spain, where the new disease was originally identified. It was difficult to estimate both the extent and the intensity of the malady. There was some discrepancy between accounts of Spanish doctors as opposed to those of foreigners. Foreigners tended to overstate the incidence of pellagra, but offered few details in order to back up that impression. There seemed to be no public policy aimed at combating pellagra. The inflexibility of agrarian and social structures in nineteenth-century Spain allowed endemic pellagra in some regions to persist up until the early twentieth century, although without reaching the scale or intensity documented elsewhere in Europe.

The incidence of pellagra on the Greek island of Corfu initially had been confined to very few cases. The rising number of cases after 1857 was attributed to corn imported around that same time from Albania and the countries of the Danube. Pretenderis Typaldos documented cases on that island. Those who supported the theory that decomposed corn caused pellagra frequently cited his work. They lacked data of similar quality on the development and late stages of the disease. Typaldos found a link between the need to import corn to the island and the accelerated expansion of commercial agriculture within the confines of a concentrated and rigid agrarian structure. This resembled the process in Yucatan. In Corfu's case, vineyards rather than hemp plantations pushed out food crops, corn ranking chief among these (Marie, 1910: 23–25; Roe, 1973: 61).

Pellagra reached alarming proportions in Italy and was the object of much public and scientific debate during the nineteenth and the early twentieth centuries. The disease originally had been identified in the eighteenth century in the provinces of north central Spain. In 1784 Gaetano Strambio calculated that pellagra affected 5 percent of the total population in those areas where it appeared and that it afflicted from between 15 and 20 percent of the population in those districts with the highest incidence of the disease (Marie, 1910: 44). There was a general consensus that the number of cases rose in both relative and absolute terms during the first three quarters of the nineteenth century. Nevertheless, statistics available for the nineteenth century suggested that Strambio's estimates were exaggerated. On the other hand, there was also certain consensus with respect to the fact that most statistics were not entirely reliable and the incidence of the disease tended to be underestimated in the first place.

The incidence of pellagra in Italy generally is thought to have grown steadily until 1871, the year in which it peaked. After that time, the rate slowed down in the most affected regions: Lombardy, Venice, and Emilia. Those were also the areas that had a more productive and developed agriculture. Southern Italy had been relatively free of the disease. Pellagra spread more rapidly there after 1871, although never reaching the virulence seen in north central Italy. In the peak year of 1871, the number of documented cases of pellagra amounted to just over 100,000. Venice was the most severely affected region, with 5.5 percent of the rural population suffering from pellagra. It was 2.7 percent in Lombardy. In that same year, in the province of San Vitus, named for the saint that protected against the malady known as St. Vitus's Dance, pellagrins made up 13.5 percent of the total population (Marie, 1910: 47). In 1910, cases recorded in all of Italy just topped 30,000. Pellagra remained a severe public health problem in Italy until the third decade of the twentieth century, despite any decreasing tendencies.

Laws aimed to combat pellagra through public policy were enacted in Italy in 1902 and in Austria in 1904. These laws accepted the Lombroso current within the Zeist school and were oriented accordingly toward eradicating human consumption of spoiled corn. The Italian laws provided for the distribution of salt at low prices, the establishment of public kitchens in affected areas, the building of asylums for pellagrins in the most advanced stages of the disease, and the construction of public desiccating ovens to dry out corn and prevent its decomposition. Additionally, they used fines and penalties to discourage the sale of spoiled corn, and established a system for the exchange of spoiled corn for lesser quantities of corn in good condition (Marie, 1910: 383–403). Everything seemed to indicate that these measures had little effect on any decrease in the incidence of pellagra, which had been declining for thirty years before the promulgation of the laws. Zeists alleged that many of these measures had been practiced well before the enactment of nationwide laws and that such changes were, indeed, responsible for the decline in the incidence of pellagra. Everything else pointed to changes in Italian economic and social conditions as the key factor in checking the expansive tendency of pellagra. Some of those changes were related to the massive emigration of Italians to the New World. Italy had some thirty-four million inhabitants in 1900. More than ten million of them abandoned Italy between 1850 and 1920, thereby alleviating some of the most severe pressure on the economy. In any case, the action of modern scientific medicine had little impact on the history of pellagra in Italy before 1930.

As the incidence of pellagra fell in Italy, it rose explosively in Romania. Authorities recorded 33,000 cases in 1901, 55,000 in 1905, and more than 100,000 in 1906 (Marie, 1910: 63). It came as no surprise that in 1907 an agrarian revolution broke out there. The revolution led to the allocation of larger plots and more pasture land to peasants. There were also improvements in what had been unbearable conditions stemming from the extreme concentration of land. The agrarian reform of 1917–19 consolidated and expanded peasant victories and stemmed the growth of pellagra. Pellagra did not disappear entirely, however, and remained a serious public health problem until World War II (Aykroyd et al., 1981).

There was not much accurate information available on pellagra in North Africa and the Middle East. Doctor Fleming Sandwith, an Englishman, was moved to write about pellagra as a result of the profound impact of his experiences with the ailment in Cairo hospitals. Sandwith's impression was that the incidence of the disease was high, especially among the peasants of Lower Egypt (Roe, 1973: 69–76). Cotton cultivation had expanded very quickly in that region of highly developed irrigated agriculture. Cotton was cultivated at the expense of other crops, especially food crops. By default, corn became the almost exclusive staple food in peasant diets. It was quite likely that severe endemic pellagra affected many other colonial populations, but information on this likelihood was confusing, unreliable, and disjointed. Such a state of affairs was entirely too typical of people without history (Wolf, 1982), people who were insignificant from the haughty perspective of the metropolises of modern industrial capitalism.

All of the European literature on pellagra associated it not only with corn consumption, but also with living in poverty in the rural countryside, with peasants. There was a lack of consensus concerning the incidence of pellagra in cities. Italian statistics measured the incidence of the disease only among the rural population, taking for granted that it did not exist in cities. In 1910, Dr. Louis Sambon, a British physician and biologist noted for his work on the tsetse fly, found cases of pellagra in Italian cities. Sambon believed that pellagra was transmitted by insects, and took on the chore of demonstrating that those city dwellers who were afflicted by pellagra had visited the countryside and surely contracted the disease there. It was very probable that the unanimous nature of opinion with respect to the correlation between rural residence and pellagra masked a whole host of prejudices deriving from the acceptance of the industrial urban millennium. This perspective saw the rural countryside only as the repository of the last vestiges of barbarism, the persistence of ignorance, and enduring brutality. Such suspicious unanimity of opinion also was a reflection of the shame felt by urban pellagrins, who avoided exhibiting any stigma of barbarism and poverty among the ranks of the civilized.

Despite the prejudice rejecting any possibility of urban-based pellagra and the air of secrecy surrounding any cases that did appear, there was no doubt that the disease affected peasants more harshly than it did city dwellers. Peasants made up the largest and poorest class in nineteenth-century southern Europe. Poverty and misery were regularly associated with pellagra in studies of the disease. According to some authorities on the subject, poverty simply predisposed peasants to pellagra. According to others, poverty was actually a causal agent of the disease. In 1817 Dr. Henry Holland declared, "The pellagra is a malady confined almost exclusively to the lower classes of the people, and chiefly to the peasants and those occupied in the labors of agriculture." In Italy Marzari asserted "that if a villager falls into poverty . . . pellagra does not fail to crown his misfortune and put an end to his miserable existence." Many similar observations could be added to this list, but perhaps it was Lalesque who best summed up the matter when he refered to pellagra in the Landes in France: "It attaches itself to poverty as the shadow to the body" (Terris, 1964: 226). Whatever role writers ascribed to poverty, they often considered poverty itself as an inevitable if not natural occurrence, owing to no specific causes. Poverty simply was the legacy of some murky past, a burden the destitute continued to bear.

The link between pellagra and corn as the chief component of the diet was duly noted in the very first descriptions of the disease. The literature on the topic constantly remarked on this association. Just as with poverty, dependence on corn very often received only a superficial treatment, peppered with adjectives rather than full-blown descriptions, as though it were an apparent and well-known evil. Some information allowed us, nevertheless, to better grasp what excessive reliance on corn in the diet meant for those areas affected by pellagra. In Italy as a whole, one could infer that daily corn consumption was anywhere between one and two pounds per person. In Melegnano, consumption stood at six pounds per week. In Ferrara, it was two pounds a day. In Friuli, inhabitants ate over eight hundred pounds of corn a year. In lower Lombardy, servants on large estates received two pounds of corn a day except in the winter, when they only received a pound and a half (Marie, 1910: 317–20). A modern nutritional survey conducted in Moldavia in 1930 fixed the average daily corn intake at just over one and a half pounds during most of the year, including winter (Aykroyd et al., 1981). This was consistent with data coming out of Italy and served to illustrate a widespread average. For the sake of contrast, one could mention that in Mexico, a country whose inhabitants received nearly 60 percent of their total caloric intake from corn, average per capita corn consumption in the early twentieth century was a little over a pound a day. That figure fell to just over ten ounces in the early 1940s. In 1960, when Mexico began to export corn, average consumption was just under a pound a day, a figure that remained unchanged until the 1980s.

The European data also showed that corn composed up to 90 percent of total food intake by weight. That is, corn very nearly constituted the predominant, exclusive component of European diets. Fresh or canned vegetables played only a very minor part in the diet of European peasants and were less than 5 percent by volume of what corn consumption was. Meat was customarily eaten only on special occasions. Overall meat consumption was very low and made up only an insignificant part of total nutrition. Dairy products were scarce and often reserved for children. The consumption of fats was also low, and they tended to be used more as flavor enhancers than as nutritional foodstuffs. Corn's preponderant role in the diet was all the more glaring during the winter months when corn mush, only barely seasoned with minute quantities of cheese or lard to give it some flavor, was on rare occasions accompanied with pickled vegetables. The men received some supplementary food during peak periods of the agricultural cycle, when the most demanding work had to be done. The women did not receive any supplementary food and, to no one's surprise, pellagra affected them at higher rates.

If those consumption patterns were accurate, there was an adequate quantity of calories and proteins in the diet. The majority of those forced to exist on such a diet never got pellagra. There were dietary imbalances and deficiencies as far as the consumption of vitamins was concerned. Subsequent research established that the specific cause of pellagra was the insufficiency or low availability of one vitamin in particular, niacin or nicotinic acid, part of the vitamin B complex. Absolute niacin levels in corn were not especially low, particularly when compared to those of other cereals. What set corn apart was its low tryptophan content. Tryptophan was an essential amino acid that was part of a complex chemical process that required a complicated balance between components and permitted the synthesis of niacin or changed into it.

Corn's almost absolute predominance in the diets of European peasants gave rise to the first Zeist current. That school of thought credited monophagy—the consumption of one sole food—as the cause of pellagra long before the specific deficiency causing pellagra was known. This Zeist current held that no single sole sufficient and complete food existed. Resorting to solid arguments and good evidence, these Zeists maintained that any diet depending overwhelmingly on only one food would bring about the appearance of associated deficiency diseases. While the theory of monophagy, even in such a vague and general form, essentially was correct, it was subsumed or marginalized by Zeist hypotheses that looked for the origin of pellagra in spoiled corn.

Problems deriving from the way corn was prepared only added to the excessive nutritional dependence on it. In almost all of southern Europe, corn was ground raw. Bread was made with corn flour or, much more frequently, a mush cooked in water: polenta or *mamaliga*. It was often the case that the mush was prepared for several days in advance and was deliberately undercooked, since peasants found it increasingly difficult to obtain fuel. The tendency for more and more cultivated lands to be put under continuous, multicrop farming made the supply of fuel from these fields highly unlikely. Woodlands were ever more distant and their owners almost always declared them off-limits to peasants. It was not uncommon for decomposed corn to be used to prepare flour. Even though it was later demonstrated that fermented corn was not toxic and did not cause pellagra, its consumption could not have been very appetizing, lesser still in the form of a parched cold mush, cooked days beforehand. Contemporary accounts conjured up images reflecting what must have been fairly common although discouraging circumstances. One must not neglect the other side of the coin, however. Corn had become a definitive part of the peasant pallet. For peasants accustomed to eating corn mush, any improvement in the diet probably meant a polenta that had been freshly prepared and well cooked, generously dressed with cheese or with butter and accompanied by meat and fresh vegetables. If peasants would have had access to better food to begin with, pellagra would not have appeared.

Corn preparation not only illustrated an inventory of shortages, but influenced the appearance and spread of pellagra. The niacin content of corn, which was no lower than that of rice, was unavailable for absorption because it was bonded in macromolecules that did not break down during the digestive process. The process of cooking corn with lime before grinding, the alkaline hydrolysis that corn underwent in the New World, apparently broke down those macromolecules in order to free them up for their assimilation as part of the niacin that corn contained. Pellagra in Yucatan probably did owe to the lack of dietary complements, but was aggravated even further by local culinary tradition. In Yucatan, already cooked corn dough was washed several times in order to obtain a whiter tortilla, also washing away the available niacin. The practice of cooking corn with lime never was passed on to Europe. In Europe, corn was treated just like Old World cereals, which were milled raw and dry. That treatment made the lack of niacin in corn even more acute than that brought about by a diet limited only in its nutritional variation. Pellagra was a disease resulting from a cluster of deficiencies. It started with a nutritional deficiency, itself a complex phenomenon, which in turn was uncovered due to the most acute deficiency: the lack of niacin in the diet.

As time passed, the specific cause of pellagra was isolated. An effective treatment was found in the second quarter of the twentieth century. Nevertheless, the clinical explanation is not enough to understand the appearance and spread of pellagra in the Old World. It is worth asking ourselves about the factors that came together in peasant existence making it possible for corn to become so overwhelmingly important and for its deficiencies to come to light with such breadth and virulence. Poverty, which is real, is not a sufficient explanation. Poverty is not abstract, but precise and varied. What follows below is a clumsy attempt to explain that poverty, poverty which perhaps was no more pronounced than that suffered by peasants and workers elsewhere in the world. The poverty endured by European peasants may have been even less rigorous in some respects, but it was just as all-encompassing for those who suffered it and paid for it with their physical well-being, with their mental health, and, ultimately, with their life.

Concentrated landownership stood out as one of the causes of poverty among pellagra-ridden peasants. In order to obtain access to land, these peasants paid onerous rents or otherwise transferred a significant share or even the bulk of their production and their labor to landowners and their agents. Traditional seigniorial rights persisted in corn's primary domain of southern and eastern Europe, even in the face of new demographic realities and new systems of agricultural production. Population increase completely upset relations between property owners and petitioners for land. More intense and demanding forms of agricultural production contributed to a severe and almost intolerable contradiction between concentrated landowner-ship and multiple arrangements for obtaining access to that land. That contra-diction played out in a context of rigid and repressive political control. The multiplication of the forms and norms to obtain access to land, the diversifica-tion of structures of dominance, exacerbated that contradiction. Such an array of alternatives made concentrated landownership highly inconsistent with prevailing technical and demographic pressures. It also created new levels of brokerage and of control that appropriated an even greater proportion of peasant surpluses. The very diversity in means to gain access to land did not resolve the exploitation of the peasantry but, rather, exacerbated it.

In nineteenth-century north central Italy, large landowners controlled more than half the land. Almost all large landowners were nobles or institu-tions who acted as absentee property owners. They rented the land out rather than directly administer agriculture and livestock operations or agribusi-nesses themselves. Few of the large landowners rented the lands directly to those who worked them. The majority rented all their land to one sole broker, the *affitatore*. That broker set up one or several agricultural and livestock operations on the best lands and managed them himself. He rented the lands that he could not or did not want to cultivate directly to smaller brokers or directly to peasants. The largest of the local property owners were very small compared to the large absentee landowners, but very large compared to the peasantry. These local landholders controlled another significant share of rural property and partially or totally rented their lands to peasants. The most important agricultural crops were not sown in compact large plots, but in units whose size was determined by the power of animal or human traction. With few exceptions, large plantations like those in Africa or the New World did not exist. The high degree of land concentration was not reflected in the size of individual agricultural parcels, but in complex mosaics formed by virtue of their shape and extension.

Depending on the quality of the land, on plant stock, on local tradition, and on the voracity of property owners or their brokers, peasants obtained access to land as sharecroppers or as laborers under agrarian contracts. Share-cropping arrangements differed in their form of rent. Rent could be paid in kind or in money. Sharecropping with payment in kind, *mezzadria*, was more common on less productive lands. Traditionally, the harvest was divided into two halves, one for the property owner and the other for the cropper, but the terms of this type of contract changed over the years, decreasing the share going to the peasant to a third or even a fourth. Land contracts for rent in cash, *affito*, were the means by which capitalist agricultural enterprises came to occupy the best lands, although the practice also spread to peasant pro-ducers. Rents, production expenses such as the high cost of draft animals, and other out-of-pocket expenses absorbed any income derived from the sale of the harvest, especially wheat, even at the best prices. In exchange, small pro-ducers would only be able to set aside just that part of the corn harvest that provided for their own consumption. There was a gradual reduction of the size of plots available for rent or sharecropping and long-term contracts were more and more difficult to obtain. These developments effectively increased those production costs that could only be paid for with money. Corn was the crop with the highest yield and the lowest price, the cheapest food. To the extent that producers were forced to depend more and more on the market for access to land and other factors of production, dietary dependence on corn increased.

Peasants also contracted to sell their labor in many forms: daily, weekly, annually, or for indefinite periods of time. They were not only remunerated with money, that at incredibly low rates of pay in any case, but also with payments in kind. The latter varied according to the type of contract. In almost all cases, payment included some sort of food ration. Beyond that, long-term contracts could include the house that the laborer lived in or a tiny plot on which to plant a kitchen garden. Shorter-term contracts could include a share of production, which implied an indirect access to land. The food ration for the most part consisted of corn, once again the cheapest alternative. This was a business decision serving to depress the price of labor. It was not uncommon for workers to receive corn rations in a deteriorated state. In some cases the workers were allowed to keep a share of the *quarantina* corn harvest. This was the corn variety with the shortest growing cycle, maturing in just forty days. It was sown in late summer, when the land otherwise was unoccupied, and the worker tended it himself. Pellagra affected agricultural workers more than croppers, which is yet another cruel irony if one considers that these same workers tended the best land in the Italian peninsula.

The corn supply was the end result of an enormous diversity in labor and agricultural contracts. The diversity of these many, varied arrangements dovetailed nicely, without creating stable sectors among the peasantry. It was not uncommon for members of the same family to find themselves in different categories. A rural proletariat, in the strict sense of that term, had not arisen, insofar as production for subsistence and reproduction rested with rural workers themselves. The peasantry, varied and subject to complex relations of exaction, carried the process of accelerated growth and agricultural modernization forward from north central Italy. Since the last third of the nineteenth century and during the course of the twentieth century, their mobilization was the decisive factor for the structural transformation of rural Italy, in a process that was at times confused and drawn out (Sereni, 1997; Marie, 1910; Messedaglia, 1927; Romani, 1957, 1963; Felloni, 1977).

If concentrated landownership were so influential in promoting the poverty in which pellagra thrived, the breaking up of such large holdings would be bound to have a salutary effect on the incidence of pellagra. This was the case in southeastern France, where the breaking up of the great estates and public works were decisive factors in controlling pellagra. In Romania, there was a peasant revolution in 1907 and agrarian reform began in 1917. What had been explosive growth in the incidence of pellagra quickly slowed down. There had been no substantive public investments in territorial improvements that might help account for the decline in infection rates. The disease did not disappear everywhere in Romania, but pellagra was confined to well-defined areas and the number of cases fell dramatically. Corn's predominant role in peasant diets of the old kingdom did not lessen, but the access to land permitted a broadening of the quality and quantity of dietary complements (Mitrany, 1930; Frundianescu and Ionescu-Sisesti 1933; Aykroyd et al., 1981).

Concentrated landownership had been linked to the presence of pellagra since the time Casal first described it. However, agrarian structure alone could not explain all the conditions that favored the expansion, the explosive growth, and the endemic persistence of the disease. Pellagra reached its maximum incidence in the richest, most important agricultural area of southern Europe during that region's fastest period of development. Pellagra was associated with disruption in the circumstances of rural living conditions, with change rather than tradition. Pellagra was a novel disease, just as the extreme poverty and the extraordinary wealth from which the affliction sprang were themselves novelties.

The introduction of corn not only changed diets, but it also exacerbated the rhythm and harshness of peasant labor. The elimination of fallow, multi-crop farming, and a summer crop, which implied even more agricultural work days, also implied more intense land use. The commercial specialization of agriculture for profit coincided with the intensification of agriculture. New crop stocks were emerging, promoting regional specializations, at the same time that a revolution in systems of rail and steam transport permitted the massive mobilization of foodstuffs on a scale heretofore unknown. The settings for the mobilization of capital and labor also changed, not only ranging over vast distances, but also changing endeavors. All the capital necessary to create an environment favorable for industrialization and urbanization in those countries without colonies emerged from the southern European countryside. Migration mobilized enormous contingents of workers, who left one trade for another, left their country, and even abandoned their continent. More than fifty million people, almost all poor, poured out of Europe in the fewer than seventy-five years between 1850 and 1925. Money, in the form of prices, capital, or salaries, was what prompted this new mobility and made it possible in the first place. The very breadth and scale of this migration was heretofore unknown. Money penetrated reaches where it previously had been completely unknown. Money created new occupations and disrupted the ways entire populations traditionally had made their living. Money circulated everywhere only to become concentrated in a few hands in undreamed of proportions. There was very little that was beyond reach of this new wealth.

The peasant economy in southern Europe became increasingly monetarized as a response to the pressures and conditions of the new social and economic order. One cannot speak of being integrated into capitalism, for capitalism was already a long-standing phenomenon. Rather, this was a new phase of capitalist integration. Monetarization did not wipe out subsistence economies, but it did limit subsistence to fewer and fewer items of production and consumption, those which were most crucial to peasant reproduction and perseverance. Corn became the essential staple in peasant life and peasant capacity to work. Corn was one means of production that capitalism, despite its power and its technology, still could not produce more cheaply.

Pellagra was, in the last analysis, a symptom of a process of fierce modernization in peripheral areas. Change was promoted in the periphery from above and from abroad in order to recreate society in accordance with an ideological model: the industrial millennium that sought to establish a homogeneous world. In order to achieve such a goal, everything in the past, all that was varied, all that was different, had to be destroyed. It was a question of an authoritarian modernization that relied on subjugation, on the concentration of power, and on exclusion in the name of progress, with the inevitable evolution of humanity toward the one true civilization. It was modernization with a vengeance that destroyed and changed everything in its path in order to preserve dominance and promote disparity. It was modernization without democracy. Pellagra was not simply a disease of poverty and deficiencies, but one of the many diseases of modernization, of development, of pro-development capitalism, to be more precise.

Pellagra disappeared in Europe in the first half of the twentieth century. For many of those who wrote about it and analyzed it, pellagra was a closed chapter in the history of medicine and society. Effective and affordable treatment did away with pellagra forever, according to many of those who studied it. We cannot be so sure that this is the case. There were troublesome indicators suggesting that the incidence of pellagra in Africa was on the rise in the late twentieth century. Other indicators suggested that conditions that first led to the expansion of pellagra were now coming to a head in the Third World. The processes associated with authoritarian and brutal modernization, belated and dependent, were unfolding there even then. Other deficiency diseases, some hauntingly familiar and others unknown before now, are emerging as a side effect of development. The increasing inability of many poor countries to feed their populations by relying on their own domestic agricultural production is not an encouraging sign. The episode with pellagra could well be unfinished, after all. I always have been a misguided prophet. I hope I continue to be one.

11

CORN IN THE UNITED STATES: BLESSING AND BANE

Pellagra appeared in the United States for the first time in the early years of the twentieth century. The new disease spread quickly and raged at epidemic proportions throughout the Old South for forty years. Corn was not new to this region. On the contrary, it was age-old. Several thousand years before European colonists and African slaves would make that region an integral part of the capitalist Western world, the indigenous peoples of the eastern seaboard grew corn as their dietary mainstay. Paradoxically, the very survival of the first European colonists and their subsequent transformation into conquerors of this new land would have been impossible without corn. Early settlers learned to grow corn just as Native Americans did and were quick to adopt corn as their main staple.

Early documents duly recorded the names of those Native Americans who taught European colonists about corn cultivation and the corn plant's many uses: Kemps and Tassore or Kinsock, Powhatans from what would later be known as Virginia in the South, and Squanto, one of the last of the Patuxets of Massachusetts in the North (Carrier, 1923: 119, 140). Surely there were many more whose names simply were lost to the historical record. In the most frequently repeated version of early contact between European settlers and Native Americans, those indigenous peoples figured as noble savages who voluntarily passed on their botanical legacy. Native American knowledge was freely given to their European successors, who were now to occupy that region as the rising stars in the next historical drama to be played out. Another, less well known but more plausible version dispenses with such romanticism. Kemps and Tassore were captives of the Virginian colonists in 1609, hostages, to be exact. The colonists described them as "the two most exact villaines in the countrie" (cited in Carrier, 1923: 119). Besides the many tasks Kemps and Tassore carried out, they were compelled to show their captors how to sow corn. The surrender of that legacy was not a voluntary act but a coercive one (Carrier, 1923: 114–37).

English colonists had attempted to cultivate some Old World cereals ever since they first disembarked in Virginia in 1607. Harvests were disappointing.

This widespread crop failure had many possible explanations. The colonists' extremely limited practical agricultural experience had to have been a factor. Colonists had been eating corn well before the first harvest of 1609. Only half of the colonists who came in this first wave survived, and corn consumption was directly responsible for that. At times they traded with the Native Americans for corn. On other occasions, the colonists simply stole corn or obtained it as regular tribute exacted by force of arms. Such practices continued well after settlers learned to grow corn for themselves. The Native Americans also learned many things, such as the use of firearms, and sustained bloody and prolonged wars with the invaders in Virginia despite the good offices of Pocahontas. She was another figure that conventional history wrapped in a romantic aura: the beautiful Native American princess who took the side of the whites, of civilization, and married one of them. It should not be forgotten, however, that the romantic interlude began when she was abducted by the colonists in 1613. She converted to Christianity, was married off to John Rolfe in 1610, and died of smallpox in England in 1617. Her gestures were short-lived and they did nothing to stave off the defeat and the ultimate elimination of the Powhatans (Fausz, 1981).

The story of Squanto was even more complicated. Squanto was a polyglot who happened to speak English and was a good deal more worldly than the Plymouth colonists. Squanto was a native of Massachusetts who had been captured by Europeans during an exploratory expedition in 1614. He was taken to Malaga, Spain, where he was sold as a slave. There he converted to Catholicism. No one knows just how Squanto got there, but he arrived in London in 1617, where he lived for two years and learned English. Squanto returned to America in 1619 as a guide and interpreter for yet another expedition. He either could not or refused to return to Europe with the expeditionaries. Squanto stayed in America, but he was a man without a people. The Patuxets had completely disappeared by then. They had fallen victim to one of the many epidemics stemming from European contact that had annihilated and dispersed his people. Other Native Americans were distrustful of him. Squanto could claim loyalty to no nation. For that reason he was the perfect contact for the Pilgrims disembarking in Plymouth in late 1620. He acted as both guide and interpreter. Squanto taught the Pilgrims how to grow corn in the spring of 1621. He introduced agricultural innovations, some probably self-inspired and others gleaned from his cosmopolitan experiences, such as burying a fish together with the corn seed to serve as fertilizer. Squanto tried to use his influence in order to act as an intermediary between the Pilgrims and the Native Americans. He died in 1624 before successfully consolidating such a role (Salisbury, 1981).

The Pilgrims were religious dissidents hailing primarily from English cities. They knew very little about farming or surviving in the wilderness. They paid a high price for that ignorance: half did not survive beyond the first winter. But the twenty acres of corn they planted under Squanto's guidance were fruitful. The same cannot be said for the five acres that they planted with European seed. In 1621 they celebrated the harvest and their survival with a ceremony of thanksgiving. The celebration was designated as a national holiday during the presidency of Abraham Lincoln. Among other things, Thanksgiving commemorated native North American plants and animals that permitted the continued presence of successive waves of Pilgrim settlers in the New World. This was a unique case in the history of the New World after the conquest, just as the historical memory of the names of the Native American teachers was unique. Perhaps the purpose of such exceptions primarily served to salve the conscience of the new masters of the land and of history.

The European colonists learned agricultural techniques from the Native Americans. The most important of these was shifting cultivation. Using nothing more than an axe, they felled trees, cleared land, and burned off the stubble to create fields. All the land that the colonists opened to farming from the time of their arrival up until the first decades of the nineteenth century had been covered with dense forests thousands of years old. The transformation of these woodlands into fields reminiscent of those in Europe that permitted the use of horse-drawn plows would have been a long and arduous process. It typically required several years, primarily due to the tremendous difficulty of uprooting large tree stumps. The colonists, without supplies and with almost no food reserves, were working against the clock. They had to grow food right away in order to survive. Traditional European crops and agricultural techniques had to be rejected as unsuitable. Indigenous techniques and plants were used instead because they allowed the land to be used immediately. These methods relied on fire in order to clear the forest and allowed for sowing crops between tree stumps without uprooting them first.

Traditional European grains, with very low yields per unit of seed, were not appropriate for hoeing in discontinuous areas. Corn, on the other had, afforded a plentiful harvest however it was planted. Four seeds were sewn in each open hole, these spaced about a yard apart, according to Native American practices. Once the seedling had sprouted, soil was formed into a mound around the stalks using a hoe. These mounds or hills grew by accretion after several years of cultivation. The seeds of other crops were planted between the hills of corn: squash, beans, sunflowers, all hailing from the Native American agricultural repertoire. Mixed farming not only increased the yield of the soil. It also fomented agricultural practices that eliminated competition from weeds. Most especially, mixed farming provided for varied and complementary food in the diet. Lands were cultivated for several years in a row until yields declined. Then these lands were abandoned in order to allow the forest to come back and replenish the soil. In the meantime, yet another field in the forest was cleared of trees.

Initially, colonists were able to use fields cleared by the Native Americans. Soon, however, they had to open their own parcels. Colonists used the same methods that Native Americans used, although with the benefit of more efficient metal implements that permitted the clearing of larger plots. With the help of draft animals they began the process of removing stumps and transforming the cleared fields into arable lands in order to allow the cultivation of Old World plants and cereals. Corn was not displaced by these European cereals. Rather, corn also came to be cultivated with the aid of the plow. Corn's high yields and its role as a mainstay of the colonial economy guaranteed that it would continue to be cultivated. The colonial economy was varied and enjoyed a high level of autonomy and self-sufficiency. Corn did not dominate the colonial economy. Rather, corn was a conspicuous part of a diversified agriculture that also included other cultivated crops, native fruits, maple syrup and maple sugar, hunting, and domestic stock raising of Old World draft and dairy animals, swine, poultry, and New World turkeys. Commercial surpluses were necessary but limited in scope, and always existed to some extent. Markets were not very well developed and little money was in circulation. An agricultural folklore arose that effectively integrated the technical and botanical repertoires of different peoples in a society that did not look kindly on integration in other, social arenas.

The colonists used corn primarily as food. Young tender corn was eaten directly off the cob. Fully mature corn was ground into flour in wooden mortars, yet another practice borrowed from indigenous tradition. At times patties made from unground corn that had been baked on the hearthstone until they were partially done were finished up by burying them in the embers to continue baking in the ashes. This was known as hominy, a Native American word. Later this same word was used for coarse mortar ground corn flour in order to differentiate it from finer flour obtained from stone grinding. A large variety of breads or patties were prepared from corn flour, sometimes called pone, another term derived from a Native American word. Corn bread

Corn in the United States

was more common than rye or wheat bread. Perhaps still more common were grits or mush, cornmeal cooked in water like European polentas, that were served mixed with milk or cream, or even molasses or lard. There was almost never a shortage of fuel in the heavily wooded lands of North America. Corn mush constantly cooked in a pot over the hearth fire, which never went out. Occasionally cornmeal was fried in butter. There was no shortage of meat, especially smoked and salted pork. Domestic swine was almost universal and in some places pigs ran wild. The diet probably was monotonous, but it would seem that food was plentiful and balanced.

Settlers made use of every part of the corn plant. The leaves, which were cut before the harvest, became fodder for the larger livestock. Corn stubble left in harvested fields became grazing fodder for domestic livestock. Corn husks were used to stuff mattresses and to weave rugs and make twine. Corncob dolls clothed in corn husk dresses were ready playthings. Hulled corncobs were used as scrapers, pipes, or fuel. Fermented corn mash became beer and whiskey. From time to time, the grain markets were not sufficiently attractive for the sale of surpluses. In times like those, surplus corn was earmarked for animal feed, and eventually transformed into milk, eggs, and meat. Pork and whiskey both were considered to be forms of concentrated corn. Pork concentrated proteins. Whiskey concentrated calories, flights of fancy, and illusions. Corn pervaded work, food, and relaxation, every conceivable aspect of day-to-day existence.

Corn was everywhere and became the organizing axis of pioneer agriculture and pioneer subsistence. Corn set precedents for the sequence and style of work and served as a bridge for the transformation of agriculture. Corn was the foundation of the household economy and allowed for the preservation of a high degree of self-sufficiency. Corn was also the basis for the realization of surpluses and participation in a wider market. Corn was the means that permitted successive waves of pioneers to settle new territories. Once the settlers had fully grasped the secrets and potential of corn, they no longer needed the Native Americans. Indigenous peoples were wiped out, scattered, or relocated as settlers penetrated even further inland.

The pioneer presence was not an isolated incident, but an uninterrupted phenomenon that lasted until the late nineteenth century. As the years went by, pioneers of one region became farmers, and new pioneers went on to settle yet another new frontier, like concentric circles. The circumference of each circle was larger and the contingent of pioneers on its outer edges more numerous. New pioneers used corn to repeat this settlement pattern time and time again, with only little variation. For nearly 300 years there was an open agricultural and territorial frontier in North America, with a corn belt on its outer fringes. The opening up of those lands was not only the result of the free mobilization of the population. Settlers did not target previously vacant lands. Rather, they moved onto lands already effectively occupied by other groups of people historically entitled to it. The driving force behind the doctrine of manifest destiny worked to subjugate and often summarily do away with those groups standing in the settlers' way.

In 1630, a little more than 4,500 European colonists lived in the English colonies of Virginia, Massachusetts, New York, and New Hampshire. The population had grown to nearly 2.8 million by the time the thirteen colonies declared their independence in 1776. Probably more than 90 percent of these people worked in agriculture. They were limited to working a narrow strip of land along the eastern seaboard that was bounded on the west by mountain chains paralleling the Atlantic coast. After Independence, settlers crossed the mountains and initiated a period of accelerated territorial expansion. In 1850, when the continental United States had nearly reached its modern dimensions, its total population had risen to over 31 million people. Nearly two-thirds of these inhabitants worked in agriculture. The majority of the population at that time still was concentrated in the eastern woodlands. Settlers had barely begun to penetrate the prairies and planes in the heart of the continent. Until 1870, more than half of the total U.S. population worked in agriculture. That percentage fell to less than 50 percent by 1880 as the total population rose to 50 million. Although the relative numbers of Americans devoted to agriculture has declined ever since, in absolute terms the number of farmers continued to grow until 1910, when it reached 32 million out of a total population of 92 million. The number of farmers remained more or less stable until 1940, when it began to fall in absolute terms. Many of those millions were pioneers up until the late nineteenth century, when the frontier disappeared.

Not all pioneers became independent owner-operators of the land they worked. As far as that was concerned, many agricultural workers never were independent pioneers or worked on land that they themselves owned. The vast agricultural and territorial frontier never was open to all. A varied and complex system of claims, appropriation, and land holdings imposed restrictions and favored concentration of land and unequal access. During the colonial period, the British Crown granted the ownership of enormous expanses of land to mercantile companies for purposes of colonization. They also ceded large holdings to the nobility, court favorites, and adventurers, especially in the middle colonies. These large properties were very difficult to occupy and afforded low returns to their absentee landowners. While these holdings eventually broke up, the landholding arrangement that came in their stead only furthered inequitable access to land. Colonial authorities also distributed lands directly. In some cases, especially in the North, these land grants were to religious communities that distributed the land in freehold among their members. In others, such as in Virginia and the southern colonies, grants were on an individual basis and free of charge. Each colonist traveling to that territory, an expensive prospect to begin with, had a right to a grant of fifty acres, and as much again for each person that he might bring along with him (Cochrane, 1979: 24–30).

Land was abundant and inexpensive for those who could get it in the colonial period, but the labor to work it was scarce. The extent of the labor shortage became self-evident and acute when tobacco emerged as a commercial crop for export to England. Tobacco was an indigenous plant grown by Native Americans. The use of tobacco spread rapidly throughout sixteenth-century Europe for its medicinal purposes and, especially, to smoke. John Rolfe, Pocahontas's husband, was believed to have been the first to cultivate tobacco in Virginia in 1612 using Native American techniques. By 1617 "the marketplace, and streets, and all other spare places" of Jamestown were sown with tobacco (cited in Carrier, 1923: 126). In 1619 the Virginia colonists could buy a European wife for 120 pounds of tobacco, one year's harvest, inasmuch as ninety young maidens of good character were brought over to Virginia (Carrier, 1923: 126). In the same year a Dutch slave trafficker sold twenty black slaves to the Virginia colonists to work in tobacco cultivation. These were the first slaves in the English North American colonies (Stampp, 1956: 18).

In the South, plantations devoted to the specialized cultivation of a single commercial export commodity arose. The southern climate favored the cultivation of plants that could not be produced in northern Europe. Cured tobacco was the most dynamic and important of these during the colonial period. Tobacco exports went from 100,000 pounds in 1620 to 102 million pounds in 1775. It was not the only commercial crop for export. Rice, introduced in South Carolina in the seventeenth century, indigo, silk, hemp, and flax also were farmed on commercial plantations prior to Independence, but on a smaller scale than tobacco. Cotton and, to a lesser extent, sugar cane became plantation crops after Independence. Until the Civil War, southern plantations were characterized by the use of servile or slave labor. Meanwhile, in the North, agricultural production was based on family and wage labor. Servile and slave labor were used only very early on and even then in a marginal way in the North. Agriculture there was primarily oriented toward domestic consumption, although it was linked to foreign trade. Ever since colonial times, North American agriculture had been bimodal in the recruitment of a workforce, a characteristic that it retained until the twentieth century. Each mode was characterized by diversity consistent with the specialization, the structure of production, and the scale favored by internal differentiation.

In the first century of the colonial period, indentured servants played a central role in the supply of plantation labor. Commercial companies recruited indentured servants from among the lowest classes in England: vagabonds, the unemployed, petty criminals, and the poor. Commercial companies transported them to the North American colonies, where they sold them to colonists and planters. In exchange, buyers acquired the unlimited services of the indentured servant for periods that varied from between four and seven years, during which time these servants received room and board but no wages. Once the contract was up, the former servants became free and poor in the new land. Although indentured servants were used as much in the North as in the South, they were concentrated especially on tobacco plantations, the most potentially profitable enterprises.

Slowly, indentured servants were replaced with African slaves. By the end of the seventeenth century, African slaves already numbered 28,000 out of a total population of a quarter-million North American colonists. That is, African slaves made up 11 percent of the population there. In the eighteenth century the pace of the introduction of slaves stepped up. North American slave traders in New England and New York played a key role in that process. The British slaving fleet controlled the slave trade for the entire eighteenth century. Its numbers were estimated at 265 vessels in 1771. Of those, about 70 belonged to North American slave traders who carried out a triangular trade with the English colonies of the Caribbean. In 1760, half a million slaves composed 20 percent of the total population (Schlebecker, 1975: 44–45). A century later, on the eve of the Civil War that would abolish slavery in 1865, the slave population amounted to nearly 4 million people, accounting for very nearly a third of the total population in the southern slave states (Stampp, 1956: chap. 1).

By the 1820s, the United States had become the country with the most slaves on the entire American continent, even though it had received only around 6 percent, or nearly 600,000, of the total number of African slaves who disembarked in the New World. Nearly half of all African slaves imported into the United States were acquired in the three decades just prior to 1807, the year the federal government of the United States prohibited participation in the slave trade. The slave population within the United States grew at high

Corn in the United States

rates when compared with growth rates for slave populations of other American slave-holding nations. This was partially explained by more favorable conditions for North American slaves (Fogel and Engerman, 1974: chap. 1). The better conditions for the North American slaves were real, but the tremendous obstacles standing in the way of manumission and restrictions on racial mixing also had a hand in the stepped-up growth of the slave population. The number of mulattos and free blacks in the United States was much less than in Brazil. Brazil received more African slaves than any American nation, but in 1825 it already had fewer slaves than the United States. The vast U.S. frontier never was open to settlement by slaves.

The tremendous impact and institutionalization of slavery in the United States during the nineteenth century coincided with the spectacular growth in cotton cultivation. In 1801, 21 million pounds of cotton were exported. By 1820, that figure rose to 128 million pounds. Southern cotton exports, almost exclusively destined for England, reached 1.8 billion pounds by 1860, almost fifteen times the amount some forty years earlier (Cochrane, 1979: 69). Cotton became the most lucrative and dynamic branch of U.S. agriculture and foreign trade. Cotton fiber was equally crucial to the Industrial Revolution in England, in which the textile industry played a leading strategic and dynamic role (Wolf, 1982: 267–95). By the middle of the nineteenth century, more than three-fourths of the cotton imported as raw material for the British textile industry came from the U.S. South. The U.S. Census of 1850 documented 2.8 million slaves in the agricultural sector. Seventy-three percent, or some 2 million slaves, lived on plantations or units devoted exclusively to the production of cotton. By comparison, only 14 percent of slaves lived on tobacco plantations, 6 percent on sugar cane plantations, and 5 percent on rice plantations (Fogel and Engerman, 1974: chap. 2).

Corn had its most important and long-lasting role as a foodstuff in the predominantly rural world of the U.S. South. In the North, urbanization and the operation of more effective and far-flung commercial systems changed diets. Urban diets came to depend less and less on corn. Rural diet became more varied, consistent with diversified agriculture and better distribution of rural income. Even on the new western frontier, where corn remained important to pioneers, new transportation routes that were opened in the second half of the nineteenth century ended dependence on corn sooner than in the past. In the South, the growth of a highly specialized agriculture only tended to reinforce the importance of corn as food. Southern plantation agriculture was associated with a very polarized distribution of income and with a large number of slaves who had virtually no monetary income at all.

Corn was the main staple of slave diets. The standard ration of corn for slaves was a peck of corn a week, or about two pounds of corn a day for adult slaves, both men and women. This was similar to amounts of corn consumed by European peasants in regions affected by pellagra. The difference was that slave rations also included half a pound of lard a day, fatty salt pork, or fat back, all rich in niacin. Corn and pork, a form of concentrated corn, were part of slave diets year-round, but they were not the only components. Slaves also received rations of salt and molasses as well as other seasonal foods, although in lesser amounts than corn and pork. The rations provided an adequate and monotonous diet, if not always a very appetizing one. Slave diets were just enough to keep the biological machines that constituted the planters' capital up and running. In general, food preparation was done by slave families themselves. They put a lot of their own time and effort into making these rations into appetizing dishes, to transforming simple nutrients into an element of human dignity, to converting cookery into a form of resistance. Slaves also obtained other foods on their own. These they gathered in the countryside—small animals or herbs—or they gleaned secretly at night from the plantation storehouses. "Roast pig is a wonderful delicacy, especially when stolen" was a popular saying among slaves (Genovese, 1976: 599). In many cases, but not always, slaves were allowed to work tiny garden plots on the little time they had to call their own. There they planted vegetables, spices or aromatic herbs, and even corn to feed their own domestic animals. Undoubtedly, many slave owners departed from the norm and distributed fewer rations than was customary. This gave slaves the justification they needed for systematic robbery from plantation stores as a legitimate form of self-defense. Pilferage as a form of protest was not always rooted in hunger but, rather, in the terms of slavery itself. Deviations from nutritional norms were neither long-lasting nor widespread. There was no evidence that slaves suffered from chronic hunger or malnutrition. As far as we know, neither pellagra nor other deficiency diseases appeared among the slave population at large. On the contrary, one of the most complex and richest culinary traditions in the United States arose among the slaves, although many of the ingredients that gave it variety depended on all the resourcefulness those same slaves could muster.

There were many explanations for the predominance of corn in slave diets. All had a grain of truth to them. Slaves had been accustomed to eating corn since long before their arrival in America. No wonder that grain was the staple food of the slave traffic. Corn took well to Africa, where it was known as couscous. Corn fit into traditional African diets effortlessly. Typically, corn mush was cooked in water and steeped in various sauces or stews. Corn had been the most common crop and staple of the diet in the U.S. South since before the arrival of African slaves in the first place. Corn thrived throughout the same region in which slavery also prospered. It could be sown and grown by the slaves themselves, at virtually no cost, with little effort, and produced yields no other cereal could rival. Robert Fogel and Stanley Engerman calculated that only 6 percent of slaves' work time on cotton plantations was devoted to the cultivation of corn for food (1974: 42). Corn did not require the best lands, which could be used instead for commercial export crops. Corn surpluses had multiple uses. Excess corn could be sold at market. It could also be converted into a concentrated form, such as pork or whiskey. These products had a greater value added and had a higher consumer demand than corn alone. Corn was, once again, the least expensive maintenance food, easy to preserve, and flexible in both its use and preparation.

In fact, corn was the leading crop in the entire South during the years of the unchallenged reign of King Cotton. Corn occupied between five and twelve times the total surface area planted in cotton in the fifteen slave states. The value of corn was almost double the value of cotton in 1849. The four most important cotton-producing states at that time were Georgia, South Carolina, Alabama, and Mississippi, together accounting for 75 percent of total domestic cotton production. Still, the surface area planted in corn was greater than acreage in cotton and despite cotton's high commercial worth, the value of corn was two-thirds that of cotton (Kemmerer, 1949). Unlike cotton, corn was everywhere. It was sown on the smallest farms in the poorest districts as well as on large cotton and tobacco plantations. While virtually all cotton was sold for cash on the commercial market, providing the engine for mercantile consumption and profit, a very considerable part of the corn crop never left the unit where it was produced. There it was used as food, whether for the hundreds of slaves on a large plantation or for a poor family in hill country, as animal feed, or processed into beer and whiskey. Despite its ubiquitous nature, the impact of corn tended to be subtle. The role corn played in self-sufficiency lessened its presence in the marketplace at the same time that it lowered costs for the production of plantation crops. King Cotton, by contrast, was a high-profile crop. Cotton favored the concentration of wealth and power and was the vehicle that transformed the plantation and landowning elites into the social and economic axis of the Old South.

The incredible concentration of wealth on large plantations at times obscured the diversity of southern slave society. Only some 385,000 white families, one in four, owned slaves. Half of these had fewer than five slaves and another fourth had fewer than ten. The planter class was identified by owner-ship of at least twenty slaves and there were fewer than 50,000 owners in this category in the entire South. Of these, fewer than 3,000 families at the very apex of the system held more than a hundred slaves. From the point of view of the slaves the distribution was reversed: only one-fourth of slaves belonged to owners with fewer than ten slaves and three-fourths to units with more than twenty slaves. The planters were a small minority among slaveholders, but owned a majority of the slaves. Nearly three-fourths of southern whites, representing a little more than half of the total population, were small agricul-turists who owned no slaves at all and relied on family labor in order to survive in precarious circumstances (Stampp, 1956: chap. 1). Despite the nu-merical supremacy of the population not directly linked to slavery, life in the Old South revolved around the institution of the slave plantation. One fig-ure contended that 1,000 families had incomes together totaling more than $50 million a year, while 660,000 families had a total income reaching only $60 million a year. That worked out to an average income of $50,000 a year for those at the very top and $90 a year for the rest. Slave families, who had no income and were considered to be just one more plantation expense, were not included in that calculation. Another estimate for the same time period set the cost of maintenance for each slave at $15 per year (Vance, 1929: 44–47).

Most southern whites formed part of a rural society that had little par-ticipation in the marketplace due to the concentration of wealth on slave plantations. Corn, as grain or concentrated in pork, was their most important food. White agriculturists perhaps depended on corn to a lesser degree than slaves, but corn was crucial to them as well. Many of the same southern dishes were eaten by slaves and whites alike, although whites perhaps had more access to a variety of complementary foods. What whites and blacks had in common was an economy in which their diets depended to a high degree on the production on their own unit, on self-sufficiency, and on restricted mar-kets. They also shared food preferences and a common culinary tradition, although access to food was never entirely egalitarian. Even wealthy planter families, whose tables were spread with the most bountiful and sumptuous foods prompting the most inspired and envious descriptions, had much in common with traditional southern cooking. This should come as no surprise, since cooking was done by slaves, true masters of their craft (Genovese, 1976: 540–50).

A whole array of profound and powerful contradictions arose between the North and the South. The North found itself in the midst of the process of industrialization. The South had an agriexporting economy. Their ideological and political differences ultimately came to revolve around the problem of slavery. Slave states and free states had been coexisting in a certain state of equilibrium. That coexistence eventually became intolerable, and in 1861 war broke out. The Civil War was bloody and prolonged, claiming more lives than any other war in U.S. history. At the conclusion of the war in 1865, slavery was abolished upon ratification of the Thirteenth Amendment to the Constitution.

The Confederate war effort disrupted much of the southern economy. The northern blockade prevented the exportation of cotton and other plantation crops during the war. Plantation production for export suffered. Large plantations and small agribusinesses dependent on slave labor remained up and running by producing foodstuffs during the conflict instead. Both armies depended on corn for the bulk of their rations, but this was especially true for southern soldiers (Bogue, 1963: chap. 7). Slaves were not drafted as Confederate soldiers, and they almost surely would have refused to serve in any case. The ultimate defeat of the South dismantled much of the antebellum regional economy. The Confederate monetary and banking systems collapsed. Abolition momentarily broke the back of southern agribusiness. The largest share of the region's capital had been invested in slaves and that was undone with their freedom. Land, another important factor of production, underwent tremendous devaluation. Land prices fell by an average of 50 percent in the first decade after the war as land values collapsed. For some years, the southern landowning elite did not command sufficient capital to produce much of anything. Production had to be restructured and came to be redirected toward food crops. Corn, the clandestine king, emerged supreme in the full light of day.

The end of the war did not bring about any serious attempts to redistribute land for the benefit of former slaves. Promises made to slaves of forty acres and a mule were gone with the political wind. With rare exception, plantations were not seized and they remained for the most part in the hands of their former owners or new buyers who maintained the tract intact. Slaves received their freedom, but they lost any sense of security they may have had. Former slaves often found themselves without any way to make a living. They gained rights in theory that in many instances were not realized in practice for many years. Former slaves acquired a new degree of mobility, but their relative position did not budge from the lowest rung on the social ladder. Very few former slaves could take full advantage of their new circumstances. The chaotic years of Reconstruction only brought further impoverishment to the majority of former slaves. Many drifted aimlessly about looking for work. Some worked as human beasts of burden, laboriously pulling plows on borrowed or abandoned land.

Demand for cotton and other southern plantation crops had not let up with the war. Buyers abroad and in the industrial North were willing to pay top dollar for cotton. Southern plantations began to rebuild in a society that had been forced to change, but in which inequality persisted. The wealthy had land but lacked capital in order to make it productive. Former slaves had their labor power, their knowledge, and nothing else. There were some failed attempts between 1865 and 1870 to continue collective work arrangements on plantations in the form of gang labor, paying an annual wage at harvest time. Instead, an alternative model gradually emerged: sharecropping (Zeichner, 1939).

Cotton plantations were rebuilt using sharecropping as their organizational axis. Those plantations promoted the extensive use of land and prolonged and perhaps exacerbated inequality. The model did not favor and even discouraged the investment of capital in land and technical improvements, often to the point of depleting the soil and eventually rendering it totally unproductive. Everyone became alienated from the land. In their role as local or absentee landlords, the new planters channeled any profits into other activities entirely. Agriculture became impoverished and lost ground technically, neither stagnating nor luring landowners and middlemen with the possibility of attractive returns. Before 1880 the production of cotton had already recovered the volume that it had reached before the war, at 5.4 million bales in 1859. Significant growth continued and by 1900 production reached 10 million bales despite low prices, or perhaps because of them. The trend continued and 11.2 million bales were produced and sold at extraordinarily high prices in 1919, the best year in the history of the New Plantation (Fite, 1979). Cotton reestablished its speculative and uncertain reign, with prices that fluctuated in proportions from one to six, with the bad years outnumbering the good.

A whole array of conditions and contracts pertained between landowners and sharecroppers. Cash tenants paid a fixed rent in money or in kind in exchange for land, a house, and fuel. They freely administered the land and disposed of the crop as they saw fit. They were independent producers distinguished and separate from other types of tenant farmers. Sharecroppers, or share tenants, received land, a house, household fuel, and a third or fourth of the amount of fertilizer used. Tenants put in their own labor, implements, work animals, forage for livestock, seed, and the remaining fertilizer. In return, sharecroppers paid out a third or a fourth of their total production. Croppers, who contributed their labor and half the fertilizer, received everything else from the owner and paid out half their total crop. In the case of the sharecroppers and croppers, the landowner was responsible for the technical direction of the productive process, for bookkeeping, and for marketing the crop. The crop itself remained as surety to guarantee the payment of rent, frequently in combination or in complicity with local merchants. In many places the cropper was not considered to be a tenant in a legal sense, but a free laborer instead. It was common for landowners to work a small part of their plantation themselves by contracting permanent and casual salaried laborers, sometimes some of their own sharecroppers (Vance, 1937; Johnson et al., 1935). The differences between salaried workers, croppers, and sharecroppers were not substantive. Rather, they reflected the age and composition of the family. The common denominator was poverty and subservience.

Another system for commercial agricultural production arose that shared certain traits with sharecropping. In the furnishing system, landowners advanced food, consumer goods, tools, and fertilizers to sharecroppers, taking the harvest as surety. It was not uncommon for large landowners to establish their own stores or commissaries on the plantation or in town. Even the smallest landowners had agreements with large landowners' commissaries or with merchants in order to prevent the flight of the harvest from the area and to share in the profits from commerce and credit. It was impossible for sharecroppers to survive until harvest time without credit. This commercial system counted on that fact. The cost of credit was very high and it was almost impossible to establish its exact impact. Merchants or furnishing landowners reaped profits from extending credit in several ways. Interest rates fluctuated at between 16 and 25 percent and were charged on the value of any advanced merchandise. Over and above the nominal interest rate there was a surcharge on items bought on credit, which was higher than or equivalent to the nominal interest rate. The surcharge and the interest together amounted to 50 percent over and above the usual retail price of the item. On top of that, the crop was purchased from the producer at prices set by the storekeeper, who received an up-front or under-the-table commission per transaction.

All this was largely irrelevant in any case because the storekeeper or the landowner kept the books as they saw fit and balanced them in their own favor. Many sharecroppers bitterly recounted how it was virtually impossible to end up in the black after selling the harvest and settling accounts. Any protests were ill-advised, since the books were done at the same time that yearly sharecropping contracts came up for renewal. Between the cost of rent and the cost of credit, even in the best of circumstances the cropper typically received scarcely a fourth of the value of production. Very often, depending on yields and prices, that share was not enough to survive. Best-case scenarios were few and far between in any case. According to one study, barely 10 percent of sharecroppers received surplus cash when the harvest was sold. Sixty-two percent miraculously broke even. The rest went into debt (Johnson et al., 1935: 12). That debt accumulated year after year, growing, setting up more profound and severe bonds of dependence and submission, eventually compelling the sharecropper to flee and start all over again somewhere else. This was not uncommon and only created a vicious circle inasmuch as the storekeeper turned around and jacked up interest rates even further in order to make up for credit losses.

The first agricultural census to count sharecroppers was carried out in 1880. The number of sharecroppers amounted to around a million and accounted for a fourth of the four million farms in the country at large. In 1900 the proportion of units operated by sharecroppers had risen to 35 percent and to 38 percent in 1920. In 1935, nearly three million tenant farmers represented 42 percent of the total of seven million farms. At that time, more than half of the rented farms were in the southern Cotton Belt, where they represented more than half the total units of production. More than two-thirds of the units that produced cotton were worked by sharecroppers. Of those, more than a third were croppers, the most dispossessed of all producers (Vance, 1929: chap. 3; 1932: chap. 8; 1937; Johnson et al., 1935). The large increases in cotton production were attributable to the absolute growth in numbers, the malignant proliferation, of sharecroppers. The typical sharecropper received between twenty-five and fifty acres of land, according to the number of family members who could work. A pattern of extensive production using family labor and a mule with stable or decreasing yields was readily repeated and multiplied. The application of fertilizer was the only innovation of any significance. The impact of fertilizers was marginal, however, only allowing for already depleted soils to continue to be worked under the same conditions or permitting marginally fertile hill country to be planted in cotton. Owners invested as little as possible and displayed virtually no vision or initiative. Cotton plantations reached their maximum expanse and most productive form thanks instead to the labor of millions of sharecroppers.

At the beginning of Reconstruction, the majority of plantation workers were freed blacks, former slaves. The system gradually incorporated impoverished whites from both the immediate vicinity of antebellum plantations and from evermore distant areas. Whites cultivated 40 percent of the cotton in

1872. By 1910, whites worked 67 percent of cotton acreage on 58 percent of the land. In 1935, southern white sharecroppers, over one million strong, easily outnumbered the 700,000 black sharecroppers. As a category, sharecroppers and their families amounted to eight million persons. While at first glance it seemed that poverty had in many ways put whites and blacks on an equal footing, differences between the groups persisted. Only 46 percent of white farmers were sharecroppers, while 77 percent of black farmers were (Vance, 1929: chap. 3). The hardships endured by the poorest members of both groups were not the same. Shared adversity did not bring the two groups any closer, yet alone make them equal. On the contrary, white and black sharecroppers developed a competitive relationship in which prejudice was exacerbated and distance between blacks and whites increased. Discrimination and segmentation were part and parcel of the New Plantation.

Corn, the clandestine king, never lost its key role in diets and trade even while its production underwent a reorganization. Corn grew on 88 percent of all agricultural units in the South, whether cultivated by sharecroppers or independent agriculturists. Cotton grew only on 73 percent of all units (Vance, 1932: 197). This meant that sharecroppers cultivated corn as well as cotton and had to surrender a part of their corn harvest to the landowner as rent. Ultimately, they surrendered virtually all their corn as a result of a credit system in which production became part of a crop lien system. As an effect of a controlled market, corn production became separated from corn's direct consumption, just as it had on the slave plantation. Sharecroppers depended on the food rations that the furnisher advanced rather than their own production. Just as happened under slavery, the rations that the furnisher advanced came to reflect a widespread norm. The standard ration of a peck of corn a week remained unchanged under the furnishing system. Not so the half pound per day of lard, which was almost halved to four pounds every two weeks for each worker in the sharecropper's family (Johnson et al., 1935: 18). Credit was not furnished in unlimited amounts. Rather, it was meted out according to crop forecasts. It was no easy matter to get any additional credit. It was necessary to solicit, to harangue, to entreat.

On the eve of the twentieth century the southern diet was characterized by the three M's: meat, meal, and molasses. It was essentially the same diet that had been fed to slaves. This diet only differed in terms of amounts and most likely the quality of individual items. The consumption of lard or fatback had been cut considerably. The availability of varied dietary complements probably also had been diminished. Commissaries and country stores had only very limited assortments of foods. Besides the three M's, about the only things that were available were refined white wheat flour and some canned goods. The selection of merchandise largely depended on the reckoning of the furnisher and what he was favorably inclined to order. At least a third of sharecroppers' total income, and often half or more, went to buy food. The variety of that food was minimal and servings were meager. It is entirely possible that the intensity and harshness of working conditions actually had grown under sharecropping arrangements. Upon the dissolution of collective labor systems, sharecropping encouraged the incorporation of less fertile and less suitable lands. Under the sharecropping system, the number of people who were poorly nourished and who even went hungry rose considerably. Traditional southern cuisine did not disappear entirely, but fewer and fewer of the people who lived in the South were able to partake of it.

It was under these circumstances that pellagra made its appearance in the southern United States at the dawn of the twentieth century. In 1906, Dr. George H. Searcy diagnosed eighty-eight patients at Mount Vernon Insane Hospital, an institution exclusively for blacks, with acute pellagra. Of those patients, fifty-six died of the malady. The diagnosis served to heighten doctors' awareness of the disease, since they were not trained to recognize pellagra. Apparently, the disease had cropped up earlier but had gone unnoticed. Pellagra began to reach alarming, near epidemic proportions. The news of pellagra in the South was the straw that broke the camel's back. The region already was infested with an epidemic of hookworm, attributed to poor sanitary conditions and poverty. This intestinal parasite entered the body through the feet, which obviously had to be bare and typically went unwashed for long periods of time. The northern press called it lazyworm and in Mexico it was known as miner's disease. A donation from the Rockefeller family financed a campaign to effectively combat it (Etheridge, 1972: 3–39). On top of everything else, the boll weevil was wreaking havoc in the cotton fields and the price of the fiber sharply fell in 1908.

The entire South came under fire for its backwardness. Not even large landowners went unscathed. In 1909 a national conference on pellagra took place in South Carolina. Corn stood accused. Mr. E. J. Watson argued for the defense in his paper "Economic Factors of the Pellagra Problem in South Carolina": "In this little State the corn crop last year had jumped to nearly 30,000,000 bushels, worth over twenty and a half millions of dollars. . . . In 1908 the nine cotton-growing states—the Carolinas, Alabama, Georgia, Mississippi, Louisiana, Texas, Oklahoma, and Arkansas, produced 561,103,000 bushels of corn—. . . . We also have thousands of head of stock hourly endangered if the indictment against King Corn be sustained. Indeed, the entire economic outlook is placed in jeopardy" (Watson, 1910: 27–28). Not only the people, but the economy and the very fabric of southern society felt the sting of the accusation against corn, their clandestine king. Corn, the only alternative in the face of cotton's uncertain future, was being threatened. Such a threat did not only loom over the South. The entire U.S. economy would have collapsed if corn were declared guilty of murder using pellagra as the fatal weapon. Systematic efforts to determine the causes behind the malady were begun accordingly.

Many scientists armed with diverse hypotheses set out to investigate this endemic disease. The clashes, the criticisms, the mockery, and, to the point where scientific grandeur permitted it, the insults abounded. Dr. Joseph Goldberger of the U.S. Public Health Service led an outstanding team of researchers. Goldberger had worked in Mexico, in Tampico and Veracruz, studying yellow fever and typhus. In 1914, Goldberger's team set off down what turned out to be the right track to uncover the causes of the disease: dietary deficiencies and lack of a balanced diet. In 1915 they demonstrated that a diet rich in milk, meat, and beans could prevent and even cure pellagra. Likewise, deficient and unbalanced diets in healthy volunteers caused them to develop the disease. One member of the team was Edgar Sydenstricker, an economist. His work on economy and nutrition showed a correlation between diets deficient in animal proteins and pellagra. Later, researchers verified the role of yeast extract, relatively poor in proteins, in the prevention and cure of the disease. This led them to single out an unknown factor that they named factor P-P, or pellagra-preventing factor, part of the vitamin B complex and not a protein. Goldberger, who died in 1929, never identified the unknown factor, which Dr. Conrad A. Elvehjem would discover in 1937: nicotinic acid, later renamed niacin. Goldberger and William F. Tanner had fleetingly looked at the reaction between niacin and tryptophan but had not pursued that line of investigation. Later researchers demonstrated the complex interaction of niacin and tryptophan after 1945. Beginning in the 1920s Goldberger's team successfully used brewer's yeast for the treatment of pellagra. As a cure, it was both readily available and inexpensive. After the late 1940s, sufficient quantities of niacin became available at affordable prices (Roe, 1973; Etheridge, 1972; Terris, 1964). Pellagra had an effective cure.

Dr. Claude H. Lavinder worked out the first figures on the incidence of pellagra in the South between 1907 and 1911 using a questionnaire sent to doctors in the region. He obtained information on 15,870 cases. The mortality rate stood at nearly 40 percent. Of those cases, 62 percent appeared among subjects from rural areas and 27 percent among city dwellers. Sixty-two percent were white and 25 percent were black. For every man affected, nearly three women were sick. The most troubling figure was the growing number of cases: from less than a thousand in 1907 to more than seven thousand in 1911. Based on the limited information available to him, Lavinder estimated those affected by pellagra in 1911 at no less than 25,000. Of those, 10,000 would die of the disease.

The high incidence of pellagra in cities was the first thing to attract the attention of U.S. researchers. Goldberger's team carried out studies on white workers in cotton mill villages in South Carolina, among whom the disease had an endemic presence. Goldberger, Sydenstricker, and other collaborators established some important facts (Terris, 1964: 111–291). Salaries industrywide were lower in the South than in the North. Real salaries had dropped in the South in recent years. Food prices were higher in the South than in the North. The supply of foodstuffs in the South was limited and commercial apparatuses were poor and onerous. That limitation was made worse by the textile industry's practice of setting up stores and commissaries, very similar to those that had been established in order to furnish sharecroppers. The southern dietary tradition of the three M's conditioned food preferences, but it was not enough to explain the glaring predisposition for pellagra. The incidence of pellagra varied inversely with family income, and especially affected the poor. The taste for meal, molasses, and meat also existed among the wealthy, but they ate those things as part of a larger diet of abundant and varied foods. The poor, who would have liked to continue eating the three M's as part of a good diet, could not afford to do so. Corn was declared innocent or perhaps only a minor accomplice to the truly guilty party: poverty.

Northerners and enlightened liberal southerners had assumed that industrialization was the self-evident and unambiguous solution to southern poverty. The conclusions reached by the researchers qualified and questioned such an assumption. Similar red flags apply today for those who hold similar assumptions with respect to the Third World. The growth of the textile industry in the South was spectacular, exceeding even planners' wildest expectations. Between 1889 and 1909 the number of spindles in the South had grown from 1.6 to 10.4 million. The number of salaried industrial workers had grown by more than 50 percent between 1899 and 1909. From 1910 to 1920, more than three-fourths of southern textile workers had spent their childhood on a farm (Sydenstricker, 1915). Despite the intensity of the rate of urbanization, the available labor issuing from the rural countryside and a defunct plantation system continued to be plentiful, even unlimited. The movement was not indicative of some irresistible attraction to industrial work, but rather it was the simple result of impoverished croppers' desire to flee the rural countryside. This circumstance encouraged industrial work characterized by low salaries and poor conditions. Only whites were hired as mill workers. This racial barrier created inequality and tension in the process of transforming agricultural workers into industrial workers. Blacks too fled the rural countryside and the South. Hundreds of thousands emigrated to northern cities. Urbanization in the South was the fastest in the United States in the early decades of the twentieth century. It was a fierce and distorted process in which a cottage industry, like some cruel caricature, was more likely to be the result than a city per se. Industrial paradise in the South was plagued with unhealthy conditions, malnutrition, and pellagra.

Not until 1927 did Goldberger's team take another look at the rural setting. Serious flooding struck the Mississippi Delta after two consecutive years of low cotton prices. Goldberger and Sydenstricker traveled throughout the region and left a dramatic testimonial of the tenant farmers' desperate situation. In that work they highlighted some cruel ironies. They pointed out that pellagra was more common where the land was the richest and most fertile. Landowners sowed cotton everywhere, not even leaving enough land for sharecroppers' cows to graze, forcing them to graze along roadsides or be sold. They confirmed that pellagra was a disease not found in nature. Land was plentiful and rich, capable of producing a variety of foods in abundant quantities. Those tilling the soil possessed the knowledge and desire to eat well off the land, free from the threat of illness, but were not allowed to. Speculators, merchants, furnishers, and landowners stood in their way. They knew perfectly well how to cure and prevent pellagra with all the certainty that science is capable of mustering, but could do nothing about it. They were powerless. They could very accurately forecast the number of deaths, but could do very little to prevent them.

Pellagra was not considered a disease that should be reported to public health officials. Perhaps this was because no one liked to belie the image of the United States as the country of prosperity, the country that had banished grievous poverty forever, the country of Manifest Destiny. So there were no reliable or precise statistics on the incidence of the malady. There were calculations that allowed estimates of the magnitude of what was called "the butterfly cast," those pellagra victims indelibly branded by poverty. In 1915 the number of the stricken was estimated at between 75,000 and 160,000. Goldberger, always conservative, was inclined toward a figure of around 100,000, a number that would remain unchanged until 1921. In the 1920s, the incidence of pellagra surged due to low cotton prices and natural disasters, such as flooding. Estimates of those affected stood at around 170,000 by 1927. It was feared that before 1930 that figure would climb to 200,000. Beginning in 1929, the Great Depression paradoxically provided some relief in the incidence of pellagra even while dealing a blow to the entire U.S. economy. Without incentives, cotton production fell off sharply. Corn and other subsistence crops replaced cotton. The tendency for the number of those affected by pellagra to climb turned the corner during the same period. In 1940 pellagra sufferers were estimated at fewer than 50,000 and their numbers fell by half over the next five years. In the 1950s pellagra became a clinical curiosity, a disease of the past (Etheridge, 1972).

It is not easy to pin down the causes of the disappearance of pellagra from the southern United States nor establish a single most significant factor. Medicinal cures played an important role. Certainly they explained decreasing mortality rates and increasing recuperation rates, but prevention could not be attributed to them. Prevention took place under normal societal conditions. There were profound changes in the South. Some of these changes were important here, such as demographic shifts between the country and the city, between the North and the South, which ended the growth of the rural population and relieved the explosive growth of cities in the South. In the 1930s, the New Deal changed the role the government played in the economy. The New Deal made public resources a powerful factor in the geographic relocation of investment and in employment, in the distribution of income. World War II, which favored the modernization of productive capacities and full employment, paradoxically consolidated many New Deal objectives. What emerged was nothing less than the complex internal and external process that turned the United States into the most powerful country in the world.

As far as everyday life is concerned, changes in the U.S. diet virtually eliminated self-sufficiency and home preparation of many foods like bread. It was not until after the Great Depression that such changes began to appear in the South. Vitamin-fortified bread provided those nutrients commonly eliminated in the refining process. Fortified bread became widespread in the 1940s and perhaps played a role in the eradication of pellagra. The "northernization" of the southern diet, resulting from changes in patterns of labor and of consumption, meant improvement in the variety and quantity of foods eaten in the South. It also meant an "indefinable loss" in the quality and flavor of the foods (Cummings, 1940: 111). The three M's no longer characterized the southern diet. Obesity, high cholesterol, heart disease, and other lifestyle diseases took the place of pellagra.

Southern agriculture also changed. It became more varied, diverse, and complex. The most outstanding example of this was the picking of cotton, the most serious bottleneck in the production process. After 1941, mechanization of labor reduced the hours of work necessary to produce a bale of cotton from 155 to 12 (Ebeling, 1979: 137). Government policies to regulate production and financial backing for southern agriculture began under Roosevelt. The New Deal eliminated the harshest restrictions prolonging the vicious circle of the plantation system. The plantation was transformed once again without ever changing hands. Diversity deposed the king—monoculture—and modernity drove the sharecropper out.

In many ways, the Old South disappeared along with pellagra. Perhaps it could be said that when economic development in the South came to a standstill, so did pellagra. When the majority of southern society was no longer brutally subjugated in the interests of furthering capital accumulation in the modern, industrialized sectors of the economy of the South, then pellagra disappeared. Pellagra was a disease born of development, a product of a type of progress that was imposed, unjust, and unequal.

12

THE ROAD TO FOOD POWER

With the early years of hunger and hardship behind them, the English colonies that would one day become the United States of America became self-sufficient in livestock and agricultural goods and demonstrated an enormous potential to produce surpluses. The scarcity of labor and lack of markets or the difficulty in gaining access to those markets emerged as the real restraints on agricultural growth, which did not come up against absolute territorial constraints until the twentieth century.

Only two extractive economic activities developed during the colonial period, and those were territorially dispersed and extensive in their use of labor. The trade in fine furs, especially beaver, had an important but limited role in colonial commerce abroad. The fur trade had a profound impact on the life and culture of Native American peoples. Native Americans were affected by the fur trade long before European colonists established direct contact with those indigenous populations. That contact was to be both hostile and enduring. (Wolf 1982: 158–94). Lumber was plentiful and also was exported beginning early on. Lumber was used to build ships, for construction in the West Indies, or for the millions of barrels used in shipping almost all the merchandise for export from the New World sugar producing colonies.

The most important and long-lasting basis for North American foreign trade was not to be furs or lumber, however. Rather, agricultural commodities and livestock were destined to become the foundation of that trade. The largest part of the United States was located geographically at the same latitudes as Europe. Those territories did not naturally complement each other, as was the case, for example, with Europe and the West Indies. Although they were at the same latitudes, there were colder winters and hotter summers in North America than in Europe. That difference and the repertoire of American crops were responsible for the fact that southern plantations that sold their products directly to the Old World metropolis prospered. First it was tobacco, the chief export product during the entire colonial era. Later, for the first century of U.S. independence, it was cotton. Other products from southern plantations such as indigo and rice, neither of them domesticated in the

New World, were added to the agriexporting repertoire without affecting the predominance of tobacco or cotton.

Diversified agriculture developed in the northern and middle colonies, where plantation crops did not thrive. The hurdles and high costs implied by the export of unprocessed agricultural and livestock products abroad before the revolution in sea transport in the nineteenth century led to other patterns for livestock surpluses and commercialized agriculture. Urbanization, more pronounced in the North than in the South, encouraged the formation of more dynamic domestic markets, although these were limited in their reach by difficulties in transport. In the North, foreign trade took on complex forms that implied the transformation of agriculture and livestock products into commodities with high prices by volume. So in northern ports, triangular trade circuits were established, derived and associated with the great triangular traffic in sugar and slaves.

The triangular trade circuit was enormously complex. Molasses imported from the West Indies played an important role in the North American leg of the triangular trade. It was the most common and affordable sweetener in the U.S. colonial diet. Molasses was distilled into rum in plants set up on the northeast Atlantic coast. Rum, in turn, served as an important vehicle for other exchanges. Rum was crucial to the fur trade as payment to native trappers. In the slave trade, rum was used to buy slaves on the African coast. Molasses was paid for in the West Indies with wheat and corn or their flours, meat or salted fish, with barrels or lumber, or with African slaves. North American ship owners hailing from northern ports, who built their own vessels, played a leading, although secondary role in the slave traffic. The triangular trade also included the direct exchange of wheat flour and biscuits with the metropolis. The goods exported from the northern and middle colonies did not surpass the value and importance of those goods issuing from southern plantations until the second half of the nineteenth century.

The two agricultural systems, the plantation and the farm, whether the self-sufficient farms of the pioneers or the commercial farms of their successors, had little in common. Corn cultivation was one of the few things they shared. The American cereal dominated North American agriculture in its geographic distribution, in area, and in volume of production in both the North and the South ever since the earliest settlement by European colonists and African slaves. In spite of being the most common and widespread crop, corn did not participate in colonial foreign trade in a significant way during the first century after Independence. Neither was it highly visible in the domestic market, precisely because it was so common and ubiquitous. In an indirect way, however, corn made it possible for agricultural and livestock surpluses to become commodities. On plantations, corn was essential to feed the workforce. On farms, it fed both workers and plow animals. Corn served as the axis of a strategy designed to meet basic needs. This, in turn, freed up commercial surpluses for the marketplace. Additionally, corn was processed into products for sale on a commercial level.

Swine and large livestock were finished with corn rations that allowed them to gain weight before going to the slaughterhouse or setting out on the difficult and debilitating road to the centers of consumption. They were the only merchandise that walked themselves to market in this isolated, sparsely populated, and extensive territory. Meats, smoked and salted, could keep for a long time packed in barrels. Their consumption was more common than that of fresh meat and, according to many, more tasty and healthy. Smoked and salted meats formed part of all diets and rations, whether intended for colonists, slaves, soldiers, or sailors. Swine and pork were thought of as concentrated corn. An empirical rule of thumb reckoned that five pounds of corn were necessary in order for a pig to gain one pound of weight on the hoof. That rule allowed producers, upon comparing prices, to make decisions on the advisability of raising pigs as opposed to selling corn. The option gave rise to the so-called corn-pig cycle that brought about the incorporation of pioneer farms into the market and their transformation into specialized commercial operations. Another rule of thumb was that four cows fed with corn generated enough waste to raise one pig, although it was rarely mentioned that the waste was in the form of the excrement of large livestock (Shannon, 1945: 165). The average consumption of meat in the United States was the highest in the world before the twentieth century, thanks to the abundance of land and corn.

Alcoholic beverages also were a form of concentrated corn. Corn-based alcohol reduced corn's volume and raised its price. Distillation transformed corn into a commodity while still providing for home consumption. Beer made with fermented corn, preferably blue or black, was a common beverage dating from the time of the earliest colonists. The distillation of corn liquor probably began in the seventeenth century. In the eighteenth century it already was known as whiskey, just like its Scotch and Irish predecessors. Whiskey was a more rustic beverage than rum. Rum was easily the most popular liquor of the colonial period. Whiskey was the preferred beverage among the frontier pioneers, who did not buy anything they did not directly produce themselves. Corn whiskey also was preferred by the many merchants who traded with Native Americans and fur trappers. Good whiskey was easy to distinguish: it was flammable. One hundred proof whiskey was 50 percent pure alcohol or fifty degrees Gay Lussac (Hardeman, 1981: chap. 13). The adulterated firewater that the Indians drank almost never ignited.

Whiskey was domestically produced for many years. It was not until 1789 that a Protestant preacher founded a commercial distillery in Kentucky. Bourbon, another term for corn whiskey, was named after the Kentucky county of the same name. In 1794, when the federal government tried to impose an excise tax on whiskey, a rebellion broke out that required the mobilization of 15,000 militiamen to contain it. The struggle between tax collectors and bootleggers did not end until well into the twentieth century. Large rum distilleries could not bear this tax burden in addition to certain changes in international trade, further contributing to the immense popularity of corn liquor in the nineteenth century. Between 1790 and 1820, when consumption was at its peak, inhabitants drank an average of just over five gallons of whiskey a year. Corn was converted to whiskey at the rate of six to one or a bushel of corn per gallon of whiskey, a conversion rate that was very similar to that of corn to pig (Hardeman, 1981: chap. 13; Walden, 1966: chap. 13). If the consumption figures for those years pertain, the amount of corn drunk by a U.S. resident was equal to that eaten by a Mexican.

The relative abundance of grains, meats, and beverages was a reflection of the abundance of land. This was not a naturally occurring fact or the product of discoveries of vast tracts of uninhabited virgin lands. It was the result of political policy, of a relation of power backed up by tremendous demographic, economic, and military resources. Territorial expansion, later protected under the doctrine of Manifest Destiny, was one of the central objectives of the policies of this fledgling republic that declared its independence in 1776. The territory that was originally recognized as composing the United States when that nation was first organized in 1783 would increase fourfold over the next seventy-five years. The United States grew by virtue of outright purchases, cessions, and treaties, always linked to wars and military confrontations: the acquisition of the Louisiana Territory in 1803 and Florida in 1819; the annexation of Texas in 1845 and the Oregon territory in 1846; the military conquest of more than half of the Mexican territory, negotiated in 1848, and the final settlement of the borders between the two countries with the Gadsden Purchase in 1853. The continental United States eventually consisted of some five million square miles stretching from coast to coast.

Still more prolonged and cruel than foreign wars were domestic wars to divest Native Americans of the territory they held and lived on. Successive military campaigns and many expeditions of armed civilians were necessary in order to vacate the lands and relocate the surviving Indians onto reservations on marginal lands. The Indian wars, from the Seminole Wars in Florida to the elimination of the Apaches in the Southwest, branded the nineteenth century and remained as one of the darkest episodes in the most spectacular agricultural development in history.

Almost three-fourths of the enormous continental expanse of the United States was at one time part of the public domain. In 1853 more than 850 million acres, almost twice the area of present-day Mexico, were part of the public domain. The transfer of land from the public to the private domain gave rise to many laws between 1784 and 1862, when Congress enacted the Homestead Act. This and subsequent homestead laws permitted settlers to occupy and later acquire title to tracts of land not to exceed 160 acres for a nominal fee. Settlers had to live on or cultivate the homestead for a stipulated period of time that could range from fourteen months to five years. This law compensated for tendencies in previous provisions that allowed land to be sold in large tracts at low prices. Those laws favored land speculation and concentration. Squatting and immense corruption contributed to the demise of this type of land allocation. The transformation of public lands to private lands was one of the most tangled episodes in U.S. history. Even though excesses were committed, this did not change the fact that an open, although selective, frontier did exist. This frontier was what permitted the extensive growth of agriculture during the nineteenth century. The transfer of lands from the public to the private domain ended as the open frontier and the century both drew to a close (Shannon, 1945: chap. 3; Cochrane, 1979: chaps. 3–5; Schlebecker, 1975: chap. 6).

The decisive factor for the spectacular growth experienced by U.S. agriculture in the nineteenth century was the absolute incorporation of additional arable land. Such extensive growth brought more and more land under cultivation without any change in the intensity of soil use. The cycle that transformed the subsistence pioneer, who rarely sold surpluses, into an agriculturist participating in the marketplace was repeated again and again over the course of the century. Corn remained important in the pioneer phase as a subsistence crop for settlers. The forefront of expansion was an ever-widening corn belt, driven forward by the pressure of an enormous swell of millions of immigrants from the Old World. More than thirty-four million immigrants, almost all from Europe, entered the United States between 1821 and 1932 (Crosby, 1972: 216). These immigrants mostly remained in the cities and few of them directly participated in pushing the frontier further west. However, their very numbers made westward expansion possible and necessary.

The transformation of the pioneer into a commercial farmer depended in great part on access to markets. The sheer size of the territorial expanse, the very thing that permitted the extensive reproduction of the units of production in the first place, meant that distance and transport costs were obstacles standing between the pioneer and the market. A true revolution in transport in the nineteenth century overcame those obstacles. Steamboats, which had navigated the Hudson River since 1807 and the Ohio and Mississippi since 1811, united the Midwest and the South. The corn-pig cycle radiated out from around this axis in order to supply markets in the South and abroad. The enormous success of the Erie Canal inspired the construction of an extensive network of canals using barges pulled by draft animals. The Erie Canal was opened in 1825 in order to unite Lake Erie with the Hudson River and, from there, with the port of New York. Its ultimate purpose was to unite the Midwest with the Atlantic coast. The Erie Canal reduced travel time between Buffalo and Albany from twenty days to just six and cut costs from one hundred dollars a ton to ten dollars.

The fastest expansion was that of the railroads. In 1830, tracks covered only twenty-three miles. Thirty years later, in 1860, tracks extended over more than 30,000 miles. After that time, rail eclipsed river transport. After the Civil War, the construction of rail lines, always by private companies that received generous land grants as incentives, accelerated even more in order to cover the vast western tracts. In 1890, the railroads covered 167,000 miles and 250,000 miles in 1910. The construction of railroads encouraged many speculative ventures, prompting the extension of the rail system beyond the country's real needs (Schlebecker, 1975: chaps. 8, 14).

Sea transport also underwent radical transformation at that time. First clippers—large, swift sailboats—then steam-powered paddle boats and, after 1843, propeller-driven steamboats shortened the ocean crossing and made it more economically viable. Regularly scheduled Atlantic crossings sailed on established routes. The domestic rail and river transport systems, together with changes in sea transport, allowed the competitive export of U.S. grain to Europe. Around 1880, U.S. grain or grain-derived exports, driven by western wheat and meat, exceeded the value of exports from southern plantations. For the decade between 1890 and 1900, Eugene Brooks (1916: 216) calculated that corn and meat produced by corn-fed livestock had a combined value that was greater than the value of all other agricultural products put together. The export capacity of U.S. agriculture, a productive power that responded to contemporary needs, rested firmly on a solid foundation of corn.

Another bottleneck for the transformation of pioneers into commercial farmers was the scarcity and high price of labor. An impressive array of mechanical inventions increased the efficiency of both human labor and that of readily available plow animals that provided the motive power on most U.S. farms. John Lane and John Deere invented plows in the 1830s that made it possible to cut and turn the dense prairie sod, otherwise impenetrable with thigh-high prairie grass. Gang plows pulled by six or more horses were devised from single plows. The invention of mechanical implements transformed almost all agricultural tasks and many of the chores associated with the processing of agricultural products. Cyrus McCormick, who invented the mechanical reaper in 1831, set up a plant to manufacture farm machinery that would become the largest nineteenth-century factory of its kind (Cochrane, 1979: chap. 10). Much agricultural machinery was designed specifically for corn cultivation. Corn responded better to mechanization than other grains because it was sown in separate rows that permitted implements to combat weeds row by row once the plants had sprouted. The abundance of land in the face of the scarcity and high price of labor made extensive agriculture attractive, precisely because extensive agriculture favored the productivity of labor over the productivity of land. This was in harsh contrast to the development of agriculture in the Old World.

As the nineteenth century drew to a close, so did the social and territorial frontier for pioneers. Specialized commercial farming came to replace the diversified farming that had been typical of the colonists and early settlers. The transformation of supply systems, together with new methods of food preservation, came to restrict subsistence farming to increasingly smaller areas. Commercial agriculture as a business undertaking became widespread. Its hegemony sentenced many types of producers to poverty: pioneers in remote areas, impoverished southern sharecroppers on former plantations, or farmers in marginal regions with depleted soils and exhausted vegetation. As the century of expansion drew to a close, the United States already had become a worldwide agricultural power.

The integration of the U.S. agricultural market brought about regional specialization there consistent with social and natural constraints. The Midwest became home to a well-defined Corn Belt. The rich black soil, level terrain, with high grasses and regular rainfall of almost forty inches a year, in what today are the states of Illinois, Iowa—the two corn giants—Ohio, Indiana, Michigan, Minnesota, Wisconsin, and Nebraska, became a privileged region for planting corn. The lands put under cultivation in that region during the nineteenth century produced yields of forty to sixty bushels per acre, 1.1 to 1.7 tons per acre, which were twice the U.S. average at that time (about 0.67 tons per acre), and were higher than those in late-twentieth-century Mexico. Since 1870 the Corn Belt has produced more that half of the total U.S. corn crop.

Corn production spread far and wide, well beyond the Corn Belt. Corn cultivation and consumption dominated in a more far-flung region loosely referred to as corn country. It included, besides the eight states of the Corn Belt proper, South Dakota, Oklahoma, Missouri, Kansas, Kentucky—adjacent to the Corn Belt and at times considered part of it—and Tennessee, whence corn got its foothold to spread throughout the Old South. Texas became one of the ten largest corn-producing states. Corn was probably the only crop cultivated in every state of the Union. Not only was corn cultivation the farthest reaching of any crop, but corn production monopolized the largest surface area of all agricultural crops in the United States.

Around 1910, the United States produced a little more than two and a half trillion bushels of corn, almost seventy million tons. That amounted to more than 1,500 pounds for each of the almost ninety-two million inhabitants recorded in the 1910 census. Such tremendous volume easily surpassed any demand for corn as food. More than half and perhaps as much as 80 percent of corn production was destined for feed for cattle, swine, and poultry. There were precedents for the use of corn for animal feed, but never prior to the nineteenth century had the largest share of corn production been used for that purpose. The history of corn up until that time had been tied directly to human nutrition. Suddenly, corn became the raw material for the production of meat and dairy products. Corn became a commodity for intermediate consumption subject to transformation. Corn's role in the United States during the twentieth century would revolve around its use as forage.

The overabundance of corn in the United States contrasted sharply with the state of the corn supply in the rest of the world and very especially with the availability of corn in those countries where it served as the primary staple food. In Mexico, for example, where corn occupied over half of all cultivated land, corn accounted for over two-thirds of inhabitants' total nutrition. Per capita production for each of Mexico's fifteen million inhabitants in 1910 would just top 300 pounds a year. This was a little more than one-fifth the U.S. average. Almost no corn remained for animal feed in countries like Mexico, in stark comparison with the enormous surpluses of corn that the United States produced for direct human consumption.

Only a modest share of the U.S. corn surplus was sold on the international market at that time. Before the middle of the twentieth century, corn exports never accounted for even 10 percent of domestic production. In the late 1890s, foreign corn sales represented around 9 percent of total domestic production, a level that would not be reached again until 1960. Corn exports went especially to wealthy European countries, where they were used as feed for livestock. Western Europe, the vanguard of rapid industrialization and modern capitalism, and especially England, had lost the capacity to produce its own basic food and grain. It was the only region in the world that regularly depended on food imports in the first half of the twentieth century. The chief grain imported by Europe was wheat and the United States became its most important vendor. Between 1901 and 1931, before the onset of the Great Depression, the United States sold an average of 178 million bushels of wheat a year to Europe, nearly 5 million tons. That amount represented more than one-fourth of U.S. domestic production of wheat and was the United States' principal agricultural export product. Average annual corn exports for those same thirty-one years was scarcely 51 million bushels a year, some 1.4 million tons. If one takes into account that average wheat prices were some 50 percent higher than corn prices, the difference between those two grains in U.S. foreign trade became greater still (Fornari, 1973: chap. 4).

In 1844, Colgate & Company began producing cornstarch. That same year, one of the company's employees, Thomas Kingsford, set up a cornstarch factory that operated until 1910. Many of the innovations that ultimately would transform modest cornstarch plants into the complex and profitable corn-refining industry took place in Kingsford's factory. In the twentieth century, the corn-refining industry came to consume between 5 and 10 percent of U.S. corn production. Hundreds of products were processed by the refining industry, from paste and oil, intrinsic corn by-products, to multiple types of sugar derived from starch. This industrial branch was based on the chemical milling of corn with alkaline. This wet grinding method was the same technique used for thousands of years to make tortillas in pre-Hispanic Mexico. In addition to this, there was the widely dispersed dry milling industry, with thousands of establishments all over the country, which transformed corn into graded flours and corn germ into oil (Walden, 1966).

Most corn, nearly three-fourths of total production, never left the farm on which it was produced, in spite of the demand of the international market, industrial consumption, and the wide market for that grain as quoted on commodities exchanges, such as the Chicago Board of Trade. On the farm, corn became pork or beef, which, in turn, were the products that appeared on the domestic and international markets. The gigantic infrastructure for transport and silage would have been insufficient to manage the immense volume of U.S. corn production, which in the first half of the twentieth century already accounted for nearly half of world production. Although it did not enter the market directly, corn was and is the most valuable commodity in U.S. agriculture, both figuratively and literally. Corn monopolized and monopolizes nearly a third of the total cultivated acreage in the United States. Within the rich and varied agriculture practiced in the United States, corn constitutes its most powerful axis, its very backbone.

Since 1895, when corn production reached two and a half trillion bushels, until the end of World War II, the total volume of U.S. corn production did not vary or grow in a substantive way. Depending on conditions, annual harvests were between two and a half and three trillion bushels, with significant deviation from this norm in only a handful of years. Average yields did not tend to vary significantly either, fluctuating between twenty-four and thirty bushels per acre or between 0.7 and 0.84 tons per acre. Total acreage devoted to corn cultivation between 1890 and 1908 wavered somewhere just over 90 million acres. This amount rose to more than 100 million acres between 1909 and 1933, a figure that remains unsurpassed. After that date, acreage in corn descended to some 80 million acres. Despite wheat's status as the leading U.S. agricultural export, that cereal occupied only a little more than half of the acreage devoted to corn. Wheat yields were half of those for corn, and production levels for wheat were only a third of those for corn (Wallace and Bressman, 1949). Like corn, wheat experienced no severe fluctuations over that same apparently stable period, which coincided with the closing of the agricultural frontier.

The apparent stability of U.S. grain production during the first half of the twentieth century was deceiving. Such stability was a reflection of an enduring system of extensive production. There had been important changes in many aspects of this system in order to maintain the continued operation and predominance of an extensive system of production. The growth of the rural population halted around 1920, when it reached 31.4 million inhabitants. Rural population growth stagnated at 30.2 million during the Great Depression and remained thereabouts until 1940. Many of those rural inhabitants were underemployed. Rural population numbers began to descend abruptly in the 1940s when unemployment nationwide disappeared as a result of the war. Rural inhabitants amounted to scarcely 23 million people by 1950. Estimates put the number of farms or units of production at 6.5 million in 1914. There were approximately 50 million agricultural units in China at this same time. The number of farms in the United States grew until 1935, when it reached 6.8 million, only to descend to 6 million in 1945. As arable land increased, the average size of farms grew: 138 acres in 1914, 143 in 1925, 155 in

1935, and 191 in 1945. The total population of the United States grew from 99 million in 1910 to 140 million in 1950, but all growth was absorbed by urban areas and nonagricultural occupations (Schlebecker, 1975: chap. 14).

A combination of many factors linked to rapid urbanization and industrial development brought an end to pressure on the land, although forms of social, economic, and, at times, racial segregation still persisted and restricted the demand for land. The average price of agricultural land grew, for example, by 70 percent between 1913 and 1920, and frustrated many an agricultural worker's aspiration to acquire his own parcel. Sharecroppers worked 25 percent of all farms in 1910. That share grew to 35 percent in 1930, only to descend to 10 percent in 1940 and rise to 22 percent in 1950 (Schlebecker, 1975: 223). These figures suggested that sharecroppers' demands for land went unmet. Between 1930 and 1950, more than a million sharecroppers were expelled from the land they worked. The idea of an unlimited frontier that could justly and universally satisfy the demand for land was dispelled. In its place was the image of a rural countryside so plagued by excess population that the only solution for its poverty and ignorance was to funnel that population elsewhere.

On farms producing cereals, the introduction of gasoline-driven tractors and associated mechanization was a crucial development in supplanting animal traction as the principal source of motive power. In 1903 Charles W. Hart and Charles H. Parr formed the first company for the manufacture of tractors on an industrial scale. Incidentally, it was they who came up with the term tractor in 1906. By 1909, thirty-one companies produced two thousand gasoline-driven tractors a year. This was the same as the number of cumbrous steam-driven tractors produced at that time. Steam-driven tractors soon would disappear forever. In 1925, more than half a million tractors were already in use. This figure rose to more than a million by 1935. In 1945, nearly two and a half million tractors were in operation, almost one for every two farms. Draft horses disappeared entirely from farms in the 1950s. By 1955, tractors numbered nearly four and a half million, almost one per farm. This number would not grow much more after that time, although the average power of each unit would rise considerably. The acquisition of trucks, combines, and other self-propelled farm machinery demonstrated the same dizzying rate of growth (Cochrane, 1979: chap. 10).

Hybrid corn seed began to be used commercially in the United States in 1933. The beginnings of hybrid seed were modest, with scarcely 0.1 percent of acreage using hybrid seed that year. Hybrids dramatically increased yields. Yields using hybrid seed could be over 100 percent higher than yields from conventional seed. The use of hybrid seed increased geometrically over a very brief period of time. In 1941, 40 percent of all corn acreage used hybrid seed. In Iowa, that figure was already over 90 percent at that time. The use of hybrid seed corn became universal by the 1950s, the first crop for which that held true. In 1926, Henry Wallace founded the first commercial company dedicated to the production and sale of hybrid seed corn. Wallace later would become U.S. secretary of agriculture and vice president under Franklin D. Roosevelt. Hybrid seed corn operations quickly became economically important enterprises, since hybrid varieties lose vigor after a single harvest and must be reproduced annually (Wallace and Bressman, 1949: chap. 3). Privately owned seed corn companies became important centers for applied agronomic research and today are the owners and beneficiaries of large genetic banks, where thousands of years of knowledge accrued by millions of producers has been deposited like germinal plasma.

The development of hybrid corn and its adoption on a commercial scale were the result of characteristics of the corn plant itself. Corn's monoecious nature—the separation of the masculine and feminine flowers—permitted directing the cross of selected varieties by detasseling or covering the masculine flowers with bags. The success of hybrid seed corn also was the result of the enormous scientific interest in the corn plant in the United States. Corn cultivation and corn improvement were the subject of copious research and numerous monographs over the course of the entire nineteenth century. Scientific agronomy in the United States never lost interest in corn like it did in Europe. This was no wonder, since corn was the most valuable agricultural plant the United States had to offer. Above and beyond the scientific research, which will not be reviewed here, was the patent enthusiasm for corn unabashedly displayed by the cultivators themselves. Beginning in the nineteenth century, corn growers formed associations that mounted exhibitions, awarded prizes, and even constructed enormous corn palaces for fairs and festivals. They organized youth groups, which still exist, that competed in yield tests. It was not uncommon for these young producers to come up with yields that were the equivalent of over 300 bushels per acre on their small test plots, or almost nine tons per acre. A culture arose around corn, a form of social relations between producers, a folklore, and even a mystique that now is dissipating due to uniformity in agricultural practices. This is yet another topic that will not be touched upon here, but that cannot be overlooked as a component of the accelerated transformation of U.S. agriculture.

Other technical changes came about only slowly before World War II, such as the use of chemical and organic fertilizers. By 1959, more than a third of all farmers still did not apply any fertilizer at all to their commercial crops. U.S. agriculture stubbornly persisted in its extensive nature, which was only possible because land was so plentiful. Since the end of the nineteenth century, pioneer subsistence agriculture had gradually disappeared. It was replaced by commercial agriculture producing for the marketplace. Still, in the first half of the twentieth century the largest part of the means of production were not acquired on the open market at all, but were supplied by the self-same farmers. Labor, motive power, seeds, organic fertilizers, all formerly had been resources that producers themselves supplied. Technical changes slowly changed that situation and promoted an increasing dependence on the market to produce. Conditions conducive to the rise and rapid growth of intensive agriculture, almost industrial in nature, were the result of these processes. The productive resources of farmers increasingly counted for less and less.

More important than technical advances were institutional changes in the role of government in the first half of the twentieth century. These changes created a powerful and complex apparatus to support agriculture. Beginning in the early twentieth century, the U.S. Department of Agriculture was transformed into a powerful scientific organization that generated knowledge and technologies, compiled information, and promoted legislation in order to regulate any manifest problems. At the same time, the USDA lacked both the intent and the capability to carry out programs and policies directly. Land grant colleges financed by their own agricultural production and other endowments were obliged to prepare scientists and agricultural technicians. These institutions also participated in generating knowledge and technologies. They, too, lacked the capacity to implement policy directly beyond the dissemination and circulation of their own recommendations. Those public institutions supplied the foundation of basic knowledge that would lead to the expansion of scientific agriculture after World War II. Almost no other public policy had such a direct bearing on agricultural production.

Access to credit was one of the most pressing demands of late-nineteenth-century agrarian movements. The limited measures of the Federal Farm Loan Act of 1916 partially fulfilled the need for the creation of a system of credit for farmers. The credit system grew slowly and erratically. Ultimately, it was insufficient to cut short the agricultural crisis that began in 1920. That downturn preceded the general crisis of 1929 by a decade and bottomed out in 1932.

What had been ineffectual government intervention in agriculture changed in the 1930s. President Roosevelt and Secretary Wallace implemented a number of agricultural policies associated with the New Deal in order to address the most profound economic crisis in U.S. history. After 1933, many laws

The Road to Food Power

and programs were aimed at decreasing production and stabilizing prices by means of subsidies. Those subsidies allowed the government to acquire and store any surpluses of the most important crops through price supports. Subsidies cheapened and improved conditions for financing agriculture. Subsidies also financed the construction of public works in rural areas: irrigation, electrification, roads, and soil conservation programs. If analyzed individually, the effect of any particular policy was uneven. As a whole, these policies acted to ensure that leading branches of agricultural production came to depend directly on government aid in order to remain viable. Producers came to count on subsidies as a crucial component of their income. The complex and intricate network of direct government subsidies or indirect subsidies in the form of price supports became one of the dynamic forces behind rapid agricultural growth after the Second World War. It remains so to this day.

The intensity and rate of the growth of U.S. agriculture sped up appreciably after World War II. This growth was in the realm of productivity and took place without bringing any new lands under cultivation. Corn production surpassed the three trillion bushel mark, 84 million tons, in 1956 and reached four trillion bushels, 112 million tons, in 1965. Ten years later, it exceeded 5.5 trillion bushels, 154 million tons, and in the early 1980s it hit 224 million tons, more than eight trillion bushels, nearly half of production worldwide. The area dedicated to corn cultivation in the United States, which was almost 80 million acres in 1945, had fallen to some 55 million acres in 1970, only to rise again to some 75 million acres in the early 1980s. Corn yields, which in 1945 were 35 bushels per acre, in the early 1980s surpassed 110 bushels, rising from just under one to just over three tons per acre over the same period.

The incredible growth of U.S. corn production could be explained by technical factors such as the genetic engineering of miraculous hybrid seed. Corn hybrids were increasingly productive and increasingly demanding as well. Chemical fertilizer became indispensable for growing grain. The rate of nitrogen application in 1970, facilitated by the commercial availability of ammonia anhydride beginning in the 1950s, was seventeen times greater than what it had been in 1945. The rate of potassium application in 1970 was nine times the rate of 1945. The disparity between the consumption of fertilizers and rising productivity was a reflection of diminishing returns in the relation between the rate of fertilization and declining agricultural yields. Chemical herbicides also became widely available after the Second World War. Herbicides and insecticides became essential elements in the agricultural production process. Machinery became bigger, more efficient, more versatile, and dramatically reduced the amount of human labor required for corn production. According to David and Marcia Pimentel (1977), the twenty-three hours of effective labor required to cultivate an acre of corn in 1945 had been reduced to just seven hours by 1975. That reduction was all the more impressive when measured per ton of harvested corn. Almost ten hours of work were required to produce a ton of corn in 1945, while in 1975 only slightly more than one hour was needed.

The rise in the productivity of U.S. agricultural labor was reflected in the marked drop in the number of people working in that sector. The number of people living on farms, calculated at 15.6 million persons in 1960, dropped to 7.5 million in 1979 and went from 8.7 percent to 3.4 percent of the total population, consistent with occupational definitions in use until 1970. The agricultural workforce was 8 million strong in 1950. By 1970 it had fallen to 3.2 million, excluding migrant agricultural labor. The number of farms continued to decrease, from a little more than four million in 1960 to a little less than three million a decade later, and to 2.7 million in 1977. The farms that disappeared were small or medium-size farms that did not have the capital to expand and compete. As a result, the average size of farms continued to grow until it reached 440 acres, 325 acres of which harvested some crop in 1974. The concentration of resources and production on the largest units increased rapidly. In 1977, less than 20 percent of farms sold more than 78 percent of all commercial crops (Cochrane, 1979: chaps. 8–9).

Strictly speaking, the term "farm" no longer reflected the character of the unit of production when applied to the largest of these. They were capitalist enterprises, businesses, complex bureaucratic organizations, sometimes the property of large corporations, livestock and agricultural factories, on which all the factors of production, and often even the administration and technical management, were acquired on the open market. U.S. farmers, defined by the use of their own labor power and that of their family, by virtue of their personal tie to the land and resources, had not disappeared, but they were no longer the leading sector in terms of total agricultural production. The era of the farmer, like the pioneer before him, was waning in the United States. The era of agribusiness had begun and already was the dominant force in U.S. agriculture.

Around 1980, U.S. corn production was just over a ton a year per inhabitant, a full 25 percent more than in 1910, despite the increase in population from 92 million to more than 220 million inhabitants over the same period. Just over 271 pounds, 13 percent, was never converted into grain at all; rather, the entire plant was ensiled to be used later as fodder. Just over half a ton, 48 percent, was used for animal feed, largely on the very same unit on which it

The Road to Food Power

was harvested. Thus, 61 percent of total corn production was consumed by domestic livestock. Twenty-six percent of U.S. production was exported, some 567 pounds per inhabitant, which was more than the amount consumed by a Mexican in a typical year. An important percentage of this volume was used for forage in the countries that imported it. Of the remaining 271 pounds, 13 percent was used as seed corn and as raw material for the U.S. refining industry and domestic dry milling.

Average domestic per capita corn consumption around 1980 was very similar to or perhaps just slightly higher than that of 1910. The radical modification of U.S. diets put corn and other cereals in a different context in 1980 than in 1910. In 1880, corn flour provided 650 calories a day per person. At around the same time, wheat supplied a little more than 1,000 calories. In 1910, the number of calories provided by the consumption of corn flour dropped to 200 and by 1960 it scarcely made up twenty-five calories a day on average per person (Bennett and Pierce, 1961). Wheat consumption did not make up for the drop in corn consumption. In 1929, U.S. inhabitants ate 177 pounds of wheat per year per person; this number fell to 107 pounds by 1975. The per capita consumption rate of potatoes, sugar, and fresh fruits also fell, as did that of almost all foods with a high carbohydrate or flour content. Meat made up the difference, especially beef, whose annual per capita consumption rose from 50 to 120 pounds between 1929 and 1975. Poultry consumption also rose from 14 to 40 pounds over the same period. The consumption of pork, on the other hand, fell from 70 pounds to 51 pounds. If the consumption of milk, eggs, butter, and lard were added to these three categories of meat, then in 1975 the average annual per capita consumption of products of animal origin was 450 pounds, a little more than a pound a day. This implied that U.S. inhabitants not only received all their proteins from animal products, but also the largest part of their calories.

The conversion of vegetable protein into animal protein was very inefficient. Twenty-seven million tons of vegetable protein suitable for human consumption were required in order to produce the 5.8 million tons of animal protein that U.S. inhabitants consumed annually. On average, 9.75 units of vegetable protein were necessary in order to produce one unit of animal protein. Beef had the lowest efficiency, with a conversion index of fifteen to one; pork was ten to one; poultry was nearly six to one; milk and eggs fluctuated between three and four to one. On top of the low biological efficiency of such conversions was the low energy efficiency of the production system. In the case of beef produced in finishing pens, one unit of energy of animal protein, a kilocalorie, required seventy units of energy from fossil fuels. In the case of milk and pork, between thirty-five and forty units of fossil fuel energy were necessary in order to obtain one unit of energy (Pimentel and Pimentel, 1977). U.S. dietary and agricultural patterns were unusual, derived from the abundance of land, which implied enormous losses of vegetable nutrients and fossil fuel, specifically oil. The unique nature of U.S. agricultural development did not prevent it from being touted as a model to be emulated worldwide.

A comparison of the per capita availability of corn in the United States between 1910 and 1980 demonstrates that corn for export was the most dynamic sector of the corn market. In fact, the per capita increase of 550 pounds in total availability over that period was practically identical to the increase in the amount of corn for export. The Second World War severely disrupted agricultural production in many parts of the world and especially in Europe. Livestock production was affected even more severely than agriculture. In the United States, on the other hand, whose territory was not a theater for armed conflict, the war acted as a stimulus to agricultural production. At the end of the war a tremendous volume of grain and foodstuffs poured into Europe in quantities higher than any ever recorded in the first half of the twentieth century. The combination of postwar demand and government-sponsored agricultural support programs stimulated U.S. cereal production more than proportionately and swelled government acquired surpluses. In 1952, stocks of agricultural products belonging to the U.S. government were worth nearly $1.3 billion. In that same year, the cost of programs to reduce production and acquire and store the surplus production was close to $1 billion (Cochrane, 1979: 140). In 1954, the volume of grains controlled by the government exceeded storage capacity. A fleet of retired freighters used during the war had to be fitted out to serve as floating granaries (Fornari, 1973: 103–5).

In that same year, 1954, the U.S. Congress approved the Agricultural Trade Development and Assistance Act, also known as Public Law 480, to use U.S. agricultural surpluses abroad and contribute to the eradication of world hunger. Title I of the law allowed the sale of foodstuffs to foreign governments on credit at attractive terms. Up until 1971, payment for the foodstuffs could be made in the national currency of the client nation and had to be spent in that country itself. Title II allowed for the outright sale or, in cases of disasters or for the purpose of foreign aid, the free donation of foodstuffs to development programs. These development programs, in turn, distributed food among the beneficiaries. Title III allowed the donation of food through charitable agencies. In 1959, Title IV was approved. It allowed the United States to loan money to countries in order to buy U.S. food with low interest loans. This program

The Road to Food Power

became known as Food for Peace and became part of Title I in the 1970s (Schlebecker, 1975: 285–86). The stated altruistic purpose of this program has been harshly criticized through the years. It has been amply demonstrated that the principal beneficiary of Food for Peace was the United States and its military, political, and economic strategic interests.

The effect of Public Law 480 on U.S. exports was felt immediately. In the first five years the program was in effect, 28 percent of U.S. grain exports received the financial support provided by the law at a cost of $5 billion. In the mid-1960s, exports under Title 1 of Public Law 480 surpassed 16.5 million tons. In 1973, when international demand already was solid and entrenched, only 3.3 million tons were exported under its auspices (Lappé and Collins, 1977: pt. 9). Other programs to encourage exports, some backed by multi-lateral organizations, were added to those of Public Law 480. All of them required some degree of subsidization. In an inevitable and perverse way, the subsides even stimulated production. In 1959, the value of agricultural products owned by the U.S. government had risen to $7.7 billion. By 1965 that amount was $6.4 billion, five times higher than in 1955. The cost of regulatory programs overseeing production and prices was $4.5 billion in 1965. The end of those programs would have led to the collapse of domestic agricultural prices on the order of between 20 and 40 percent (Cochrane, 1979: 140).

These programs expanded the world market and disrupted its norms in order to provide an outlet for surpluses produced by the United States, the largest and most powerful exporter in the world. As a counterpoint, U.S. agriculture became dependent on the international market. Between 1967 and 1979 U.S. agricultural exports grew 125 percent and their value rose from $6.8 billion to $32 billion, which represented around a fifth of total exports world-wide. A third of cultivated acreage in the United States was dedicated to export crops. According to statistics gathered by the Food and Agriculture Organization of the United Nations, in 1981 the United States produced 20.2 percent of the world cereal supply, but was responsible for 48.2 percent of grain exports worldwide. In that same year, 75 percent of wheat, 57.4 percent of rice, 20.2 percent of barley, and 26.3 percent of corn raised in the United States was sold on the international market. Almost half of foodstuff exported worldwide came from the United States, which was the behemoth behind food power. The new reality was just this: global interdependence in a grow-ing world market for foodstuffs became a central factor, at times the domi-nant one, in agricultural development in the second half of the twentieth century. Some authors called that new reality the world farm.

In more than one sense, corn was the crop with the leading role in the international phase of U.S. agriculture and remained its backbone. In the early 1980s, the value of U.S. corn exports practically caught up with that of wheat, which had been the most important agricultural export product for almost a hundred years. In terms of volume, corn had already surpassed wheat some years before. In 1981, the United States exported 49.7 million tons of wheat and 60.5 million tons of corn. These two crops combined made up 85 percent of U.S. cereal sales for export. But while wheat exports represented 48.2 percent of international trade in that grain, corn exports represented 69.1 percent. Wheat exports accounted for more than three-fourths of domestic production, while corn exports scarcely accounted for 26.3 percent. In 1945—taking that year as a starting point—wheat exports were close to 11 million tons and represented 35.2 percent of domestic production. In that same year, nearly 855,000 tons of corn was exported, which represented scarcely 1.2 percent of domestic production. In 1970, 14 million tons of corn was exported, scarcely 12.2 percent of domestic production, as against 19 million tons of wheat that accounted for almost half of domestic production of that grain (Fornari, 1973: chaps 5–6). These data demonstrate the incredible dynamism with which corn responded to the opening of, and at times the violation of, the international market. Only the cultivation of soy beans, introduced relatively recently into the United States, responded as rapidly as corn and became a leading export crop in the 1970s. There are serious questions about the potential for growth for soy beans. It has been nearly impossible to substantially increase soy bean yields, which between 1959 and 1973 grew less than 30 percent. Corn yields, by comparison, rose by 140 percent over the same period.

Corn yields also outdistanced any rise in wheat yields. Wheat production increased from 17.3 to 38.1 bushels per acre, from just under half a ton to just over one ton per acre, an increase of 120 percent from 1945 to 1981. While these figures are not insignificant, they paled when compared to corn production, which grew by 220 percent over the same period. The comparative advantage in price, which in the first half of the twentieth century favored wheat by up to 50 percent, fell in the second half of the century to fluctuate at around a 25 percent price advantage. While wheat remained chiefly a product for human consumption, corn entered the world market with a dual purpose: as a foodstuff for the poor and as forage for the rich. For this reason corn was able to surmount the inelasticity of demand typically associated with cereals. In wealthy countries corn was linked to the transformation of diets in general, similar to the process that took place in the United States. In poor countries, however, corn was associated only with the transformation of diets of the

The Road to Food Power

wealthy living there. Another factor was the strong pressure applied by U.S. producers. These and many other factors contribute to the explanations of the dynamic response of corn in the modern era. In the last analysis, what stand out are corn's flexibility, its malleability, and its bountiful harvest in the face of the continued exigencies and demands of humankind. This is true whether corn is produced by gigantic agribusinesses or by staunch peasants.

The long sequence of events that transformed the productive potential of U.S. agriculture into historic reality, into power and effective supremacy, are frequently related as an ever progressive and triumphant tale, especially its last chapter: scientific agriculture and accelerated development, the universe of knowledge about nature and people. Such a triumphant tale is simplistic. It ignores the successive structural and cyclical crises that have profoundly affected the development of U.S. agriculture. Contradiction and crises are constituent parts of any gains and at times explain them better than conventional wisdom. That triumphal tale, seen through the eyes of the winners, is also false. Not because the winners have not been victorious, but because they ignore the price of that victory and the fate of those left in its wake. Those who lost have not been stifled or eradicated. They endure as a part of history, as a lesson to be learned, and as problems that must be solved. The triumphal version of U.S. agricultural might, growing ever stronger daily, neglects the fact that it also is fragile and vulnerable.

It is not possible to summarize here all the contradictions in U.S. agricultural development. I will only highlight some of the problems that I perceive without trying to analyze them, only raise them as topics for reflection, as topics of interest to the reader and certainly to this author. The history of U.S. agriculture is a process of accumulation with very different and increasingly accelerated rhythms. It is also a history of inequality, of exclusion, and of subjugation. Each process created its own marginal groups, but did not completely wipe them out. Those groups persist. The Native North Americans endure, the first to be dispossessed, stubbornly clinging to their identity and at the same time segregated by it. The rural poor remain in depressed areas. In 1965 there were fourteen million rural poor, who never were reached by programs to eradicate poverty. There are the urban poor, many of them descendants of farm workers or black sharecroppers who left or were expelled from the rural South after 1945. Migratory agricultural labor survives under conditions marked by low salaries and unstable employment, many of them illegal immigrants. It is they who make those crops economically viable that do not lend themselves to mechanization at competitive costs. There are the hungry, attended to since 1965 by the food stamp program, itself ironic in a country that could not possibly eat all the food it produces. Today it is the farmers who are most affected by inequality and concentration of land and wealth. Bankers have expelled them from their lands by the thousands and those who remain face fierce competition from large corporations, whether agribusinesses or financial or commercial giants. Marginalization threatens the American farmer, the most outstanding product of the U.S. democratic ideal.

On another level, the efficiency of U.S. agriculture and the food it supplies are not absolutes. Neither can productivity grow in an unlimited fashion, contrary to the image that is sometimes cultivated. The high efficiency of U.S. agriculture is a fact, but it refers to valuable resources and restricted processes that often disturb the equilibrium between other factors and that are not always able to be replicated. The abundance of unirrigated arable land of good quality and low risk—scarcely 10 percent of the cultivated surface in the United States requires irrigation—is the general precondition. In that context, the highest efficiency is fundamentally located in the productivity of labor. As a function of that high productivity, mechanization has developed that allows workers to cover more area. The condition in which there is scarcity of labor is not generally obvious. It can even be asked if such a condition pertains in the United States any longer, if we compare the number of unemployed to the number employed in the agricultural sector. Abundant land and a highly productive limited workforce are a reflection of the large size of productive units. The highly unusual conditions prevailing in U.S. agriculture would be difficult to replicate.

Mechanization combined with intense fertilization implies increasingly higher rates of energy consumption, so high that nowadays the return is largely negative: more fuel, almost all inorganic, is invested in harvested foodstuffs. When the next step of the food chain is taken—the conversion of agricultural crops into meats—the negative return increases even further. If to this is added the fact that three times more energy is spent on distribution and procurement than in agriculture and livestock production per se, the negative return skyrockets. In terms of energy, U.S. agriculture is the least efficient in the world. David and Marcia Pimentel (1977) calculated that if the entire world population were fed a U.S. diet and foodstuffs produced under conditions identical to those in the United States, total world oil reserves would run out in only thirteen years when used for agricultural purposes alone. In U.S. agriculture, corn cultivation enjoys the highest energy efficiency of any crop, but even that has declined over the last few years (National Research Council, 1975: chap. 6). The high productivity of labor at the expense of low energy efficiency pertains, and only when the availability of fossil fuels is unlimited and their prices low.

Events after 1973 have demonstrated that such a precondition is by no means guaranteed. This will be even less the case in the future.

The productivity of soils and labor have been subjected to the ultimate criterion of efficiency in recent U.S. history: profitability. Agricultural capital must be profitable, just as is the case in any other business. Complex efficiency pertains in agriculture. That is because, despite any strictly economic objectives, agriculture is conditioned by factors that are not strictly economic. It is naive to think that profits are not affected by forces and events that have no price or market. Frontier land was not purchased. It was forcibly expropriated. Markets were opened by cannonades or with trinkets, not by comparative advantage. Slaves, before having a price, had to be captured. Miraculous hybrid seed that did not have a patent or a purchase price was derived from collective knowledge thousands of years old. Scientific research and its discoveries cannot always be subjected to cost-benefit analysis. Governments, their policies, subsidies that allow production without demand, taxes, customs, and tariffs, even altruistic donations of food, intervene in the profit margin of capital, in its ability to grow and reproduce. Cost and price, comparative advantage, are subject to and exist in combination with relations of power and dominance that do not always have a price tag, that at times cannot even be quantified. They are intermingled in order to establish a way to measure efficiency in pursuing a real but elusive resource: capital.

Absolute efficiency measured in terms of the profitability of capital generates its own paradoxes. The history of U.S. agriculture is replete with them. Some are preserved as bitter memories, such as when food reserves were destroyed while many went hungry at the height of the Great Depression. Badly eroded lands formed a dust bowl stretching for thousands of miles in the interest of speculating in an attractive although fleeting market. Profit, when managed by gigantic corporations operating under the logic of economies of scale and worldwide power brokers, often is not consistent with the common sense and the practical experience of a farmer. There are cases in which large corporations enter into agriculture specifically intending to lose money. This would be poor business strategy from any point of view other than the intricate labyrinth of tax breaks. Ronald Reagan, before he became president, boasted of not paying taxes on income thanks to losses on his California ranch. There are some who affirm that the large territorial scale of agribusiness has surpassed the optimal limits for its suitable technical operation (Robbins, 1974). It is likely that U.S. agriculture, offering the highest rate of return for ubiquitous capital, is entering into a phase of decreasing agricultural efficiency. It is also likely that the system of food procurement has become irrational in terms of the management of foodstuffs, but not as a source of profit.

Not even the maximization of profit is a widespread or common condition, although it may be dominant. The majority of agricultural producers in the world, peasants, do not operate on the basis of maximizing profit on invested capital. They do not have capital; they never have had capital. Peasants maximize production and often pursue the same objectives as agribusiness, but they do so for reasons of their own. They want to eat. They need a livelihood. They are willing to work longer or for better pay. They are determined to prevail as a group, as a culture, or as a civilization. They seek to improve their material conditions of life. Peasants seek to improve their direction, their role in history, as part of a plan or as a step toward Utopia. Standards for efficiency that contradict standards more typically associated with capital derive from here. Peasants see many of the effects of growth generated by profit as senseless. For that reason Native Americans did not understand why the whites treated the land as their enemy, as an object of conquest and submission. On many occasions, U.S. agriculture overcame the land by destroying it entirely or drastically reducing its natural complexity in order to convert it into an inert resource, one quite suitable for accounting purposes, but often quite unsuitable for life itself.

: 13 :

THE SYNDROME OF INEQUALITY:
THE WORLD MARKET

In 1972 the Soviet Union bought more than 30 million tons of grain on the world market, the largest commercial transaction of its kind in history. The operation became legendary. It was dubbed "The Great Grain Robbery" because of the great secrecy in which it was carried out and because of the intervention of multiple murky intermediaries. The largest part of the grain came from the United States, where it had been acquired at very low prices thanks to U.S. subsidies. The U.S. government was surprised, or at least feigned surprise, at the operation. In any case, after that purchase the price of grain on the world market went through the roof.

There had been a certain parity on the world market since 1960 between the price of oil and the price of wheat: about two dollars per barrel of crude or per bushel of grain. Wheat prices were the first to upset the equilibrium, surging after 1972. For a few months in 1973, a bushel of wheat was worth two barrels of crude oil. The response of the oil market was swift and retaliatory. When wheat sold for the record price of six dollars a bushel in 1974, a barrel of crude oil already had hit eight dollars. The breach was to grow even wider. A new stage had begun that signaled a restructuring of international economic relations. One of the many affects of this restructuring was an overabundance of money in world financial markets. This surplus money found its way to Third World countries in the form of loans, sometimes to buy food and fuel, and very especially to Latin America. Today we all are living in the midst of the crisis derived from this new stage that creates an uncertain future for everyone and an almost intolerable present for poor countries.

The Soviet purchase of 1972 only made sense of and hastened an already existing tendency in the world market for food. This tendency was characterized by growing demand and diminished stockpiles. The Soviet purchase remained, in anecdotal form and in collective memory, as the immediate cause behind the breach of a certain equilibrium between supply and demand during the 1960s. As the gap between the cost of crude oil and grain widened, irate U.S. farmers increasingly resented the high prices of fuel and fertilizers. They demanded that their government restore the old equilibrium with a one-to-one exchange of a bushel of grain for a barrel of oil. That equilibrium had been decisively upset, and not because of the Soviet purchase, although the entry of that country into the world market at that precise moment did signal the global scale of food dependence. Around that same time the concept of food security arose and became widespread. Food became a powerful resource in political strategy, in international negotiation, and in alignments into geopolitical blocs.

Here it is worth stopping and taking a look at a case in which corn, our protagonist, had an important and recurring role. After World War II, a triumphant Soviet Union emerged as the second most powerful country in the world. Heavy industry and advanced technology were the basis of that power, even though the high-tech part of that country's industry was concentrated in strategic areas linked to military might. Consumption demands of the population at large were for the most part ignored. Soviet industrial and military strength had an important Achilles heel: its agriculture, which was rigid and not very productive. The brutal submission of the peasantry and their production to accelerated industrialization through forced collectivization in the 1930s occasioned famines and human losses in the millions. Agriculture became impoverished as a result. Adding to the lack of resources was the large proportion of the population in the agricultural sector. Even as late as the 1980s, about a third of the Soviet Union's total population was in the agricultural sector. The combination of a scarcity of resources, a large population, and a bureaucratic, authoritarian, and vertical state administration generated a rigid structure that checked the growth of agriculture. Agriculture was unable to satisfy demand. At times it was not even capable of providing the bare necessities of agrarian producers, let alone Soviet society as a whole.

Stalin largely neglected the matter. After his death, the agricultural problem became a priority. Nikita Khrushchev, Secretary of the Communist Party between 1953 and 1964, led the agricultural reform movement. Khrushchev was the only top Soviet leader with peasant roots, having spent his childhood and early youth in Ukraine. When Khrushchev came to office, the food situation in the Soviet Union was poor, but not intolerable. The Soviet Union was self-sufficient, with an average consumption of around 3,000 calories per inhabitant per day. This was considered satisfactory and ranked among the highest in the world. The Soviet diet was less than adequate in terms of its very poor component of nutrition of animal origin and in the continued practice of rationing and frequent food supply problems. Animal protein accounted

The Syndrome of Inequality

for scarcely a fifth of the total calories in Soviet diets, a figure that was much lower than in other highly industrialized nations. Food supply problems resulted in severe scarcity and in black markets. The diet in the Soviet Union was satisfactory in cereals and other basic foodstuffs, but not in meats. Agricultural production had to rise significantly in order to address this issue, due to the inefficient rate of transformation of plants into foods of animal origin.

Khrushchev and Soviet officials designed a two-pronged strategy. The first consisted of the cultivation of New Lands. These were located in marginal subhumid regions and were subject to great climatic risks. Enormous extensions of virgin and idle lands would be devoted to the extensive mechanized cultivation of wheat. The production of wheat on the New Lands would shift that crop from more productive and suitable agricultural areas. Forage crops would be cultivated in those areas instead, in order to increase livestock production, the second front of action. To that end and to excess, Secretary Khrushchev promoted corn cultivation. He was corn's most devoted enthusiast, perhaps familiar with that plant from his childhood in the Ukraine.

In his speeches and working trips, Khrushchev frequently remarked on the importance and high yield of corn in the United States. He even vowed that, thanks to corn, the Soviet Union would surpass U.S. milk and meat production levels sometime during the 1960s. On a trip to the United States in 1959, Khrushchev was invited to Iowa to tour Roswell Garst's corn and hog farm. Garst had been associated with Henry Wallace, former secretary of agriculture and vice president during the Roosevelt administrations, a reminder that the world revolving around corn was small indeed. Khrushchev's tremendous enthusiasm for the American cereal only became greater after that visit. Khrushchev was so insistent about planting corn, bringing it up at every opportunity, that the urban bureaucracy gave him the contemptuous nickname *kukuruznik*, or "corn-nik," an unflattering reminder of the Soviet leader's peasant roots.

The *kukuruznik's* strategy seemed to be successful until 1958. Planting on the New Lands increased cultivated surface area in the Soviet Union by one-fourth. Livestock numbers rose by one third. Agricultural production was half again what it had been in 1953. Nearly 150 million acres of New Lands had been cultivated. Average yields on this new acreage, however, were disappointing at less than half a ton per acre. Only between 44 and 66 million tons of wheat were harvested on them, according to official statistics provided by the former Soviet Union. Such a yield left much to be desired. In 1954 a total of 10.6 million acres of seed corn was sown, almost all in the south of the country. In early 1955 the Central Committee of the Communist Party accepted Khrushchev's proposal to increase corn cultivation to more than 69 million acres by 1960. If the Soviet Union could meet this goal, it would mean that the area committed to Soviet corn cultivation would be comparable to the area sown in corn in the United States. The surface area sown with corn in the Soviet Union climbed to 44 million acres in 1955 and to over 59 million acres the following year. The Soviets met their goal in 1960 and exceeded it by the early sixties. By 1962, the Soviets cultivated over 84 million acres of corn, more than was sown in the United States (Volin, 1970: 327–45; Anderson, 1967; Strauss, 1969: chap. 7).

The spectacular growth in yields achieved since 1953 came to a standstill after 1958. Despite increasing agricultural acreage, by 1962 total agricultural production was scarcely 7 percent higher than what it had been in 1958. Levels were even less when measured in per capita terms. Stagnation was all the more apparent when compared with the 1959–65 plan that proposed a total growth of 70 percent over that same period. A catastrophic drought in 1963, culminating several years of unfavorable climatic conditions, forced the Soviet Union to import more than 13 million tons of wheat from abroad. In 1964, Khrushchev was deposed. Khrushchev's agricultural policy was a leading factor, if not the determinant one, for unseating him. His successors promptly suspended Khrushchev's plans for Soviet agriculture.

Unfavorable climatic conditions only exposed the shortcomings of Khrushchev's agricultural policy sooner than otherwise might have been expected. Other factors altogether were ultimately responsible for the failure of Khrushchev's agricultural strategies. After 1960, it became impossible to pay for the cultivation of the New Lands, where yields had been disappointing. Any additional investment in cultivation there would outpace net earnings going to the state. The cultivation of the New Lands was not abandoned altogether, despite the gamble and ruinous cost it implied. In the case of corn, the indiscriminate planting of corn had taken place at the expense of other, better adapted crops. Corn had been sown on climatically unsuitable lands. Corn only grew to full maturity for use as grain on 20 percent of the lands devoted to its cultivation. Corn was harvested green on fully 80 percent of that crop's acreage due to the early Soviet winters. That corn was ensiled or used as fresh forage. Corn yields for forage were disappointing at between five and six tons per acre. Yields for corn as grain also were much less than expected. Between 1958 and 1962 yields were less than a ton per acre, according to official Soviet statistics. U.S. estimates put Soviet corn yields at seven-tenths of a ton per acre. After 1960, even Khrushchev himself warned that it was risky to insist on planting corn where conditions suitable for cultivating it did not pertain.

The Syndrome of Inequality

After Khrushchev's removal from office, harsh assessments dictated that in light of corn's poor showing, the corn craze had to be abandoned (Volin, 1970: 327–45).

Other factors also shed light on the failure of Khrushchev's corn program. Corn cultivation was imposed from above, ordered, without consultation with or the participation of producers themselves. Bureaucrats and administrators promoted corn cultivation not because of any inherent advantages therein, but in order to please their superiors, to show party discipline. In the context of such an authoritarian structure, corn was planted indiscriminately, independent of geography, terrain, and peasant preferences. There was virtually no use of chemical fertilizers, their production having been very low in the Soviet Union. Also absent were any other technical resources that encouraged high yields in large-scale agricultural operations. To this must be added the general unavailability of suitable seed. Seed was not adapted to regional conditions, much less specifically produced as hybrid designer varieties. At one time, the Soviet Union had an important contingent of first-class agricultural scientists, some of them experts on corn. They were marginalized by Stalin and Trofim Lysenko. Lysenko was president of the Lenin Academy of Agricultural Sciences after the Second World War. He rejected genetic theories on hereditary traits. Lysenko denounced the work associated with Gregor Mendel, August Weismann, and Thomas Morgan as reactionary. That is, Lysenko condemned precisely those theories crucial to the development of hybrids. Lysenko not only had controversial opinions, such as denying the existence of genes and chromosomes. He also had the bureaucratic power to repress research that did not tow his rigid ideological line and even to punish any researchers who did not agree with his positions. Lysenko almost single-handedly destroyed agronomic research in the Soviet Union at that time. Khrushchev's program lacked Lysenko's active support. Curiously, Khrushchev the agricultural crusader left the power of Lysenko the Stalinist untouched. Khrushchev's corn program, although a reasonable strategy, smacked of improvisation, voluntarism, and excessive elasticity from its very beginnings. Imposed from above in a bureaucratic and authoritarian way, it taxed even the incredible flexibility of corn.

Khrushchev's successors were not any more fortunate, although they were more cautious and conservative in their agricultural policies. Less than ten years after deposing Khrushchev, they had to resort to the purchase of the century: more than 30 million tons of grain on the world market. The immediate cause was once again an exceptionally poor harvest due to bad weather. Still, the purchase was indicative of the persistence of structural problems. Soviet production as a game of chance was a reflection not only of geography but of the vulnerability and rigidity in the organization of agricultural production and of the passive subordination of peasant producers. In 1975, the governments of the Soviet Union and the United States signed an agreement for the sale of up to 8.8 million tons of U.S. grain annually. Since then and until the dissolution of the Soviet Union, that country systematically depended on cereal imports.

In 1980, 1981, and 1982 the Soviet Union imported an annual average of over 42 million tons of cereals, which represented over 22 percent of its domestic production. The Soviet Union had lost its dietary self-sufficiency. Ironically, corn was a major component of those imports, accounting for a full 30 percent or an annual average of 11.7 million tons between 1981 and 1983. Soviet corn production had not significantly risen since Khrushchev's time. Imports almost equaled domestic production between 1981 and 1983: something more than 12.5 million tons annually, on average. Yields, averaging 1.2 tons per acre according to Soviet figures, had increased by just 17 percent with respect to averages for the period 1961 to 1965. The information on corn production for forage and ensiled corn, which was not very clear, suggested a retreat with respect to the Khrushchev era. The *kukuruznik's* strategy using corn as the means to improve the animal protein component of Soviet diets ultimately held sway, except that corn had to be imported rather than be produced domestically. This postponed confronting the rigid structures that weighed heavily on Soviet farmers and peasants. Such a confrontation could not be avoided indefinitely, however, and would emerge as one of the most severe problems of that once great industrial and military power. Agricultural production once again came to have a very high priority during the reformist regime of Gorbachev.

It is time to return to the main discussion. The Soviet purchase of over 30 million tons of grain in 1972 revealed the profound changes that had taken place in the world market for food. Before the Second World War, all the world's most important agricultural regions, except for Western Europe, not only were self-sufficient, but were net exporters of cereals as well. The world grain market was relatively small, something over 27.5 million tons a year between 1934 and 1938. The role of the United States as an exporter, due to effects of the Great Depression, accounted for scarcely a fifth of this total. Eastern European countries played a similar role in the world grain market. Latin America, with 36 percent of the total, was the most important cereal exporting region in the world at that time. Argentina dominated Latin American cereal exports. The biggest buyer, almost the sole buyer, was Western Europe, the world's most highly industrialized region. Wheat was the most important grain in the world market, representing just less than half the total volume of world cereal markets. Corn followed, with a little more than a third, with rice making up not even a tenth of the total volume (Brown, 1971). For the most part, the five large multinational companies that controlled the world market regulated supply and demand. That supply and demand, in turn, more or less defined prices and quotas.

This panorama changed drastically after World War II. The size of the world cereal market grew at an accelerated pace. The nearly 30 million tons that were traded on average each year between 1948 and 1952 rose to over 49 million in 1960 and to more than 103 million in 1972. Wheat increased its share as the most important commodity. From 1948 to 1952 wheat represented three-fourths of the cereals on the world market and still made up nearly two-thirds of that market in 1970. The United States became the world's most important exporter. From 1954 to 1956, U.S. exports of more than 14 million tons per annum represented over 40 percent of the total worldwide. Between 1960 and 1962, U.S. production made up half of total exports and in 1971–72 U.S. grain for export accounted for three-fourths of the world market. Its supremacy was indisputable. In 1972, besides the United States, only Canada, Australia, New Zealand, and Argentina were major exporters of cereals.

The other side of this equation was that all the most important agricultural regions of the world, except for northern North America, Australia, and New Zealand, were no longer self-sufficient and became net importers of cereals. After 1966, Asia was the most important importing region. By 1972, Western Europe, safeguarded by strong protectionist tariffs and subsidies to stimulate agricultural production, had been able to reduce imports by some 20 percent compared to the period 1948–52. At that juncture, Western Europe accounted for just 20 percent of the international market for cereal imports. Japan acquired a proportion almost equal to that of Western Europe, and the Soviet Union and the countries of Eastern Europe bought as much again. Poor or underdeveloped countries, all of which had been self-sufficient or net export-ers of grain before the war, purchased almost 40 percent of the grain on the world market after the war. If we consider not only the balance between lead-ing agricultural regions, but also interregional commerce, then poor coun-tries became the most important clients on the world cereal market. Food dependence became a chronic and widespread phenomenon in many Third World countries.

U.S. strategic interests distorted the new structure of the world market for food. All the pure laws of mercantile exchange—price levels, supply and de-mand, comparative advantages—which never were "pure" in any case, all were affected. Large, enormous subsidies worked to channel U.S. surplus production toward international markets. One of the most flagrant examples of this, the Agricultural Trade Development and Assistance Act of 1954, came about under the humanitarian guise of food aid. Countries deemed to be friendly to the United States, especially poor countries, were made an offer they could not refuse.

The governments of poor countries found it increasingly difficult to feed explosively growing urban populations at prices they could afford. Traditional peasant agriculture and existing transport and supply systems were simply inadequate. These governments, with few exceptions, adopted plans for derived industrialization: manufactured goods were produced domestically rather than being imported. Affordable food was a crucial prerequisite for such models. Complicating the situation was a rising middle class, usually composed primarily of public servants, that demanded to live, or rather to consume, in the same way as their U.S. counterparts. U.S aid typically was in the form of affordably priced grain, payable in nonconvertible national currency, that is, money that had to be spent in the country itself. Such offers were enormously attractive and it became virtually impossible for governments that lacked the ability and the will to confront the agrarian problem to resist them.

The offer of U.S. aid was not always consistent with the dietary practices of recipient nations. The Dominican Republic and Nigeria were good examples. Geographic conditions in these countries did not favor wheat cultivation and traditional diets typically did not include wheat. Those countries still received aid for the purchase of this grain. They had to set up mills in order to grind flour, bakeries to bake bread, and even actively promote wheat consumption in a population unaccustomed to it. Further, they provided their own subsidies, in addition to those they received from the United States, consistent with their own political economic agendas in order to make the consumption of wheat bread attractive in urban areas, especially among the middle class. They generally were successful, something those nations occasionally came to regret.

In other cases, such as Colombia, a country that always had been self-sufficient in wheat, grain prices under U.S. aid programs were lower than domestic prices. The lower price of imported wheat permitted the lessening or elimination of subsidies for the politically pivotal urban population. Wheat imports discouraged domestic production. Wheat producers, faced with low prices, reorganized agricultural production around milk and meat, especially in the agriculturally privileged region of the plain of Bogota. By the 1980s,

Colombia imported more than a million tons of wheat per year despite possessing all the conditions to be self-sufficient in wheat and even to export that commodity.

Such accounts are not intended to be some sort of singular collection of anecdotes and worst-case scenarios. Rather, they are intended to illustrate the general process of substituting imported food for existing foods or foods that potentially could be produced domestically, to the point that those imported foodstuffs make up a significant proportion of national food supplies. These proportions were significant by virtue of their cost, volume, and an inexorable tendency to rise. Those levels of imported food often became even more significant because they became impossible to forego. Recipient nations did not possess the time and level of investment that on short notice would have been necessary to make up for any lapse in imported foods. A curious paradox came to mark countries receiving food aid in the 1950s and 1960s. Domestic production in those countries was geared toward the substitution of industrial imports, while basic foodstuffs, much more critical and significant in terms of sovereignty than automobiles or refrigerators, were imported. In 1971, a year before the Soviet purchase would forever change the international grain market, the purchases of foodstuffs by poor countries represented half of all imports worldwide. This was even more significant for the United States, leading the way in promoting food aid and subsidies: in 1968, poor countries purchased 78 percent of all U.S. exports.

U.S. food aid was of such magnitude that it determined the size and many of the conditions under which the world market operated. Between 1956 and 1960, U.S. aid accounted for nearly 32 percent of the world wheat market, a level that rose to nearly 36 percent between 1961 and 1965. This aid accounted for approximately 70 percent of all U.S. wheat exports. In 1965, U.S. wheat exports under Titles I and II of the Agricultural Trade Development and Assistance Act of 1954 peaked at 16.5 million tons. Since that date, the volume of food delivered as aid began a gradual and then an accelerated decline (Brown and Eckholm, 1974). In great measure the bill's objective had been met. It had generated an entirely new market, whether by introducing the consumption of wheat or by displacing existing domestic production.

The majority of the U.S. food aid under the Agricultural Trade Development and Assistance Act of 1954 was in the form of wheat and wheat flour. Corn, rice, and sorghum together never made up more than 10 percent of the aid before 1965. Not until 1973 and 1974, when aid was a mere 10 to 20 percent of what it had been at its high point in 1965, did other cereals reach a volume similar to that of wheat as components of U.S. aid. The overwhelming preference for wheat as part of U.S. aid was a logical process. Wheat had been the most important U.S. agricultural product for export during the entire first half of the twentieth century. The U.S. share of the international market for wheat was a historical given. U.S. wheat production for export had monopolized at least a third of total U.S. output for nearly a quarter of a century. This did not happen, or happened to a lesser degree, with other cereals. Pressure from domestic producers and international demand found common ground in wheat, and the aid served a dual purpose. Surpluses of other traditional U.S. export products, like cotton, could not be channeled as aid. The government had to resort to aggressive dumping in order to place these surpluses in the markets in the first quarter century after the war. If the U.S. preference for wheat was natural, its effects on many poor Third World countries that neither customarily consumed wheat nor could produce it were not. The United States created a growing and irreversible dependence on an essential good for which there was no substitute. The perversity of food dependence inspired critics to take a second look at those generous donors of food aid. Could food dependency have been a Machiavellian objective? Such a suspicion has not been fully put to rest.

In any case, it was clear that the United States had the most to gain from food aid, economically as well as politically and strategically. This has been argued and demonstrated in different ways. The clearest evidence of this was seen in the reduction of U.S. food aid as U.S. surplus production found a natural demand, independent of lucrative offers of foreign aid. U.S. food aid reached its high point in 1965. The United States provided almost 95 percent of food aid worldwide that year. That figure fell to 63 percent by 1973. Over the same period, the value in dollars of U.S. aid fell almost by half. More than 17 million tons of grain were exported in 1965 under the aegis of the Agricultural Trade Development and Assistance Act of 1954. That amount fell to just over 2 million tons by 1973. These figures, while indicative, do not illustrate the true magnitude of the reduction in U.S. food aid. Since 1971, just one year before the sharp price rise provoked by the Soviet purchase, grain purchases under the Agricultural Trade Development and Assistance Act of 1954 could no longer be paid in nonconvertible national currency. Rather, their cost had to be paid in foreign currency. In 1973, of the $730 million that the United States set aside for food aid, nearly 60 percent was in the form of loans. Worse yet, for a limited period beginning in 1970, nearly half of all food aid went to just two countries: Vietnam and Cambodia. Food aid was, in effect, part of the military supply line in a war defending U.S. interests. Mission accomplished, U.S. food aid ran out.

Food aid, contrary to its stated humanitarian purpose, always was restricted by virtue of strategic interests. The United States always selected friendly countries that could receive food aid using the most blatant political criteria in terms of a global confrontation with the Soviet Union. The United States considered Yugoslavia, Socialist and independent, to be friendly, so that country received food aid—but not the democratically elected Socialist government of Salvador Allende in Chile. The food aid that the United States extended to Allende's predecessors was denied, only to be renewed with undreamed-of generosity after the bloody military coup d'état by Augusto Pinochet. Cambodia and Vietnam, while in the midst of a U.S. war, were clear examples of the political content of the food aid and the price governments had to pay for it. Similarly dramatic cases could be added in which food aid brazenly served U.S. political and strategic interests. They all illustrated a general agenda in the distribution of food aid. Such is the reality of power. Perhaps it would be naive to expect otherwise. There should be no surprises when the altruistic and humanitarian motives with which food aid was sent and in which that aid remains steeped are rejected. Under the guise of aid, a weapon of subjugation was born: food power.

The 1972 crisis in the world grain market brought out some other little-known facts, such as the hegemony of the five large multinational companies that dominated world trade in cereals: Cargill Inc., the largest, and Continental Grain Company, both based in the United States; Louis-Dreyfus of Paris; André of Switzerland; and Bunge Corporation, the most ubiquitous, with offices in Argentina, Brazil, and Spain. There is no precise information on the volume of grains and other foods distributed by these companies. If we consider that only two, Cargill and Continental, handle half of U.S. agricultural exports, and with it at least a fourth of the world market, an estimate of at least half the world market seems reasonable, although even that estimate could be conservative. These companies even distributed U.S. food aid, since the government only gave the loans to the countries in order to acquire the cereals, which effectively were sold and delivered by commercial enterprises, to which recipient governments paid cash.

That was not the only return the five large multinationals had from U.S. aid. When loan payments were made in nonconvertible national currency, a portion was devoted to financing the establishment of foreign businesses in recipient nations. The expansion of U.S. multinationals coincided in time and was not unconnected to food aid. Large multinational corporations established subsidiaries that specialized in almost everything, including milling the wheat received as aid. Continental Grain Company did this in Zaire, Ecuador,

Puerto Rico, and Venezuela (Dinham and Hines, 1984: 141–42; Burbach and Flynn, 1980: 261). If the strategic design of the food aid was that of the U.S. government, its implementation was left to private parties, among which the five large multinationals loomed large. From the point of view of the recipients of the aid, the distinction between Continental or Cargill and the government of the United States of America was not always clear and never definitive. The same sort of thing quite likely happened at the donor end. The overlap between private and public interests in U.S. aid was quite substantial and, at times, indistinguishable.

This is not the place to relate the fascinating story of those five large multinational companies, as told by Dan Morgan (1980). Nevertheless, one aspect of these companies is relevant here. The five large cereal companies are a prototype of multinational capitalism in the late twentieth century, with perhaps one exception: they are totally controlled and principally owned by only seven families that went into the cereal business in the second half of the nineteenth century. They are doubly private: by their very nature and objectives and because their shares are not openly traded on the stock market. They are family businesses that qualify by virtue of their operating volume and their profits as among the largest corporations in the world. Perhaps because of that and because of the nature of the business in which they participate, in which secrets, confidential and privileged information, can result in catastrophic losses or monumental gains, they operate in an almost clandestine nature. Before 1972, when they all participated in the sale of the century to the Soviet Union, very few people were even aware of their existence, much less their power, their subsidiaries, and their ubiquity.

Despite their family character, the five are typical commercial multinational corporations and are at the service of international capital. They produce nothing directly, not even the goods that constitute their main product line. The cereals and food that they market are cultivated by millions and millions of producers all over the world, from peasants and *latifundistas* to farmers and agricultural enterprises, and it is they who assume all the risk. But many of the services necessary in order to produce and sell grain in the marketplace are in the power of or under the control of these companies: granaries, transport, warehouses and docks, communication networks, and so on. A good part of the financing and subsidies that enable a harvest to appear on the consumer's dinner table also are under the control of these very same grain dealers, as at times are the very means of its transformation. The products do not carry their trademark, but they have their invisible brand. They are global firms that buy and sell simultaneously in the Punjab, the province of Buenos Aires, or the state of Iowa. They are international by virtue of the scope and nature of their business and assume many faces, names, and organizational forms along the way. They would appear to have all the trappings of pure capital, contingent and even supranational.

The mobility and ubiquity of the large cereal companies seems to confirm this image of capital without borders. Continental, founded in France, now has its headquarters in the United States. Its central office is said to be found wherever the Fribourgs are. Bunge, originally established in Holland and Argentina, does not even have a main headquarters, and its central offices— especially after the Montoneros kidnapped members of the owner dynasty, the brothers Marcos and Leonora Born, in Argentina in 1975—are divided between Brazil, Argentina, and Spain. The five multinational companies pretend to be ideologically neutral and claimed to support *détente* between the United States and the Soviet Union, when they were their two best clients: one as seller and the other as buyer. Their personnel come from all over the world. No one better than they illustrates the internationalization of capital that is often spoken of as a novel and recent phenomenon.

The world grain market, arena of operations for the five large multinational corporations, is not neutral, nor does it feign neutrality. The magnitude of supply and demand, prices and the conditions of sale, such as vetoes, embargoes, and restrictions, are the result of political decisions. U.S. strategic planning and U.S.-led alliances and blocs define the vagaries of the world cereal market. On the other hand, it is domestic agricultural policy, the management of protectionist tariffs, public subsidies, and programs to stimulate or depress production, that provoke surpluses or shortfalls and affect the course of foreign policy. There is no place for neutrality or a lack of national loyalties. The five large family corporations that manage world trade become sales representatives on the one hand and advocates of domestic policy, of states and of nations, on the other. In this capacity they have acted as objective tools of U.S. food policy. They behave like domestic capital in that respect, even though they may be the best example of capital that is pure, cosmopolitan, and supranational. The expansion of capital and the interests of capital over and above the interests of individual states is a dubious affirmation or one that requires many qualifications. Capital, since it began to exercise its dominance, is immersed in the contradiction between its expansive force— which takes it to heights and interests that do not respect borders and at times are almost worldwide—and its political course, its national character. The contradiction is not resolved, perhaps because it has no resolution, just as the dispute over the predominance of the economy over politics or vice versa has no resolution. It is not the determination of one abstraction over another that is important, but their interaction and concrete combination in the social processes of history.

The 1972 crisis marked a definitive change in the nature of the world cereal market. A situation of chronic plenty, of the search for or creation of clients for perennial surpluses, suddenly became a situation of scarcity, of production insufficient to meet demand: a market of sellers caught up in the accelerated growth of prices. That change also implied an ominous reduction in world food security. The latter is measured by an index that adds the existence of grains in the exporting countries at the moment just prior to harvest and the productive potential of any idle land in the United States under government subsidy, even though it would require twelve to eighteen months in order to harvest a crop. In 1974 the U.S. government withdrew all subsidies paid to farmers to leave land idle, and that component of the reserve disappeared. That year, food reserves in exporting countries amounted to just under 100 million tons, which would cover only twenty-six days of consumption worldwide. Never before 1973 had the reserve been less than fifty days, and it was almost always more than sixty days. In the case of a major climatic catastrophe in the most important agricultural regions, food reserves would be insufficient (Brown and Eckholm, 1974: chap. 5). The threat of hunger, one of the four horsemen of the apocalypse, seemed close at hand.

Many pessimistic forecasts and even some tragic predictions were made. The gloomy Malthusian prophecy concerning the winner in the race between population and food seemed about to materialize. Anguish about overpopulation and with it programs to check population growth resurfaced with renewed vigor. A clearer consciousness also arose about dependence and food power concentrated in only a handful of countries. Food production acquired, publicly and universally, a political content that previously had been clandestine or hidden. Surpluses, once considered a problem, flagrantly emerged as a powerful and lethal weapon, a new stick used to terrify or thrash potential clients. Some blows were suffered through ill-advised and abrupt unilateral decisions. The United States restricted the exportation of soy beans and other forage crops in 1973 in order to combat domestic inflation. In Asia, soy beans were not used for forage. Rather, they were an essential foodstuff. Thailand, Brazil, and Argentina also imposed controls on food exports. The conspiratorial and secret nature of the international grain market became even more intense. The threat of hunger no longer depended just on uncontrollable natural phenomena, but also and especially on political decisions.

The apocalyptic prophecies did not come to pass. By 1977, stockpiles of more than 188 million tons, the highest in history, would provide the world with a food supply for two full months. In 1987, reserves were at 440 million tons. Illusory abundance and irrationality reappeared as never before. The European Economic Community considered destroying 22 million tons of meat, butter, and grain because it cost $4 billion a year to store them. In 1986, the United States spent $6 billion to export corn that was worth barely $2 billion. Saudi Arabia began to export wheat, with growers receiving prices four times higher than prices paid for imported wheat. At the same time, according to calculations by the World Bank, 730 million human beings, 15 percent of the human race, were not consuming sufficient calories, not to mention protein, for leading a normal working life. In Latin America it was estimated that 150 million persons lived in poverty, as opposed to 60 million in 1970. The new abundance reflected an incredible increase in production. It also reflected the appearance of extraordinary government subsidies to the agricultural sectors in powerful countries. Such subsidies were often justified by new doctrines of food security. One figure put the amount of subsidies for agricultural production and trade in developed capitalist countries at $120 billion a year. That figure was more than double the total amount sold on the world cereal market.

Such abundance cannot continue unchecked indefinitely. Dire predictions forecast that the Soviet Union would not meet the objective of producing 255 million tons of cereals in 1987. The U.S. Department of Agriculture estimated the Soviet harvest that year at just 215 million tons, over 40 million tons less than projections. Representatives of the five large cereal companies will be hatching new conspiracies for yet another sale of the century. Abundance is the unilateral reflection of an incredibly unequal growth, not only between Socialist countries and capitalist ones, but especially between the poor and the rich, countries and people. Such extreme inequality takes place in a framework of increasing interdependence, in which surpluses are expressed as insufficiencies on one side and as shortages on the other. The overproduction of food is a problem of the rich, not of the world as a whole. Inequality is the collective problem.

The world market for food outpaced population growth in the years since the 1972 crisis. Between 1981 and 1983 an average of 250 million tons of cereals was exported per year, almost one and a half times the level of the late 1970s. That increase reflected the combined effect of demographic growth and changes in eating habits among affluent classes or countries and clearly pointed to the rapid increase of food dependence. The world grain market had acquired not only a new dimension but also a new structure in which the quotas between its participants had been changed. These changes could shed some light on new linkages and dependencies, on the phenomena of increasing inequality.

The United States continued to hold its position as the largest exporter and most powerful force in the world market. Its superiority had moderated in some respects but continued undisputed. Between 1981 and 1983 it exported on average 115 million tons of grain annually, a little less than half the total. With that volume it was the world's leading supplier of wheat, with over 45 percent of the total worldwide, and of corn, with just under 70 percent. It ranked second in rice exports, very close behind Thailand, with just over 21 percent of the world total. For barley, the United States placed fourth in world exports, after Canada, France, and England. The United States was first in the export of soy beans besides, a market in which the only other country of any importance was Brazil, and even its performance lagged far behind. The proportional relationship between the cereals exported by the United States had changed over the years. Wheat production had annual averages of 47 million tons between 1981 and 1983. Wheat bowed to corn in terms of volume, which reached nearly 56 million tons, with very comparable values for both grains. If the importance of soy beans were included, U.S. crops sold as forage, not necessarily purchased for this purpose by dealers, were already more important than food crops as exports. Corn's incredible growth as a commodity for export was the most outstanding phenomenon since the crisis of 1972.

In an inevitable and logical way, changes in U.S. foreign trade were reflected in the world market. Wheat, which continued to be the leading cereal, represented a little less than half of the world market for cereals between 1981 and 1983. Corn, on the other hand, had moved up to third place. Rice scarcely represented 5 percent and barley only 8 percent. According to data from the United Nations Food and Agriculture Organization (FAO), the developed countries, Socialist and capitalist alike, excluding trade between them, remained net exporters of some 77 million tons of cereals a year between 1981 and 1983. Any generalizations ascribing the role of importers of raw materials and food to wealthy countries remained well off the mark. When this figure was broken down, developed capitalist countries improved their position even further, with a favorable balance of trade of almost 128 million tons a year, with a value approaching $18 billion. The developed Socialist countries, on the contrary, reported a trade deficit in cereals of over 48 million tons a year. Many poor or developing countries were net importers of almost 85 million tons of cereals a year between 1981 and 1983, with a value of nearly

The Syndrome of Inequality

$15.5 billion. Demand was most important in those countries, the result and the reflection of food dependence.

The high price of wheat meant that it continued to be almost exclusively a food crop. Trends in wheat sales, therefore, more acutely reflected the seriousness of food dependence. Western Europe had been the main buyer of wheat for more than a century. Then it changed course and between 1981 and 1983 became a net exporter of more than 11 million tons of wheat a year. This development was a result of Western Europe's protectionist agricultural policies and programs to stimulate production. The Soviet Union, a long-standing exporter, also did an about-face. It became the most important individual buyer in the world market with average purchases of 22 million tons per year between 1981 and 1983. The largest market for wheat was in poor countries with a market economy. As a group, they acquired almost 44 million tons of wheat a year, dispersed between dozens of client nations. Wheat could not be produced in many of them and in others it stopped being grown with the advent of food aid. For these countries, the impossibility of acquiring wheat on the world market or, more precisely, from the United States, resulted in serious supply problems and in some cases even famine. For wealthy large wheat exporting nations, such a wonderful market was the ideal place for magnificent business and political alignments.

The international market for rice was more restricted, despite the fact that rice, almost exclusively a food crop, was nearly as important as wheat. This was because the biggest consumers of rice were self-sufficient. Japan maintained one of the strictest protectionist agricultural policies and a system of subsidies that guaranteed self-sufficiency. China, the world's largest producer, also remained self-sufficient and increasingly produced for export. Even so, between 1981 and 1983 an average of nearly 14 million tons of rice were traded on the world market. Poor countries were net importers of 1.8 million tons. If Thailand is excluded, that deficit rose to 5 million tons and rivaled that of the developed countries. African countries stood out as the most important market among the poor countries with 3.3 million tons. Barley was traded mostly between the countries in the northern hemisphere, where two-thirds of the world market for that grain circulated. Once again, poor countries saw a net deficit of almost 5 million tons, with the countries of the Near East accounting for just over 4 million tons.

Corn's increasing importance on the world cereal market illustrated one of the factors responsible for its growth: the growing prevalence of meat-based diets among the wealthy countries and moneyed classes. Between 1981 and 1983 an average of over 80 million tons of corn per year were exported. Its most important single buyer was Japan, with nearly 15.4 million tons annually. The Soviet Union, with 11.7 million tons a year, was in second place. Western Europe, as a region, imported 22.7 million tons but exported 6.7, which meant that it was still left with a deficit of nearly 16 million tons annually. With the exception of France, which exported more than 2 million tons a year, historically leading corn producers had become corn importers. Spain, which had for the most part paid little attention to corn production, imported 5.5 million tons a year between 1981 and 1983. Portugal, which stepped up wheat cultivation at the expense of acreage in corn over the course of the twentieth century, purchased 2.8 million tons annually. Italy acquired as much again. Together, the developed countries bought a little more than two-thirds of the corn on the world market in order to devote it to meat production. The commercial grain market for forage crops was the most dynamic branch of the world cereal market, growing more rapidly than all the rest. It would account for the most growth in the near future, while important sectors reoriented their diets toward the consumption of foods of animal origin. That tendency, however, could not continue indefinitely.

Corn also illustrated the extent of food dependence in poor countries. Between 1981 and 1983, poor countries imported 12 million tons of corn a year. China acquired nearly 5 million tons, where it was used exclusively for food. African countries bought nearly 4.4 million tons per year as food for its population. Mexico, corn's country of origin, figured as an important importer, with 2.9 million tons per year destined to be used for urban food supplies. In lesser proportions, many other countries also bought corn in significant quantities for food. Corn's importance as a foodstuff continued to grow in the 1980s, although the volume of corn as a food crop grew more slowly than the volume of corn intended for forage.

A little play with the figures can give an idea of the magnitude of dietary dependence in poor countries of the Third World. In wealthy countries, average annual per capita consumption of cereals is more than a ton. Less than a fifth of that amount is consumed directly, while the rest is incorporated into the diet as products of animal origin. In poor countries, the average consumption is around 440 pounds. This is one-fifth of what it is in the wealthy countries and nearly all of this amount is incorporated into the diet directly as food. Imports by poor countries between 1981 and 1983 represent the total consumption of cereals by 345 million persons, something akin to the entire population of Latin America. If we suppose that each ton of imported grain complements the feeding of twenty-five persons instead of fully satisfying five, that dependence would grow to include more than 1.7 billion persons,

The Syndrome of Inequality

who would suffer severe hardship if they had to forego that consumption. In such a hypothetical case we are talking about a little more than one-third of the human race. If we leave out the Chinese, who are self-sufficient, and the Hindus, who seemingly have managed to achieve self-sufficiency, all the inhabitants of the Third World included in such a scenario would end up depending on imports for one-fifth of their basic food supply. That proportion is sufficient to transform chronic hunger into unbearable hunger in the Third World. Once again, the apocalypse draws near.

Such international number games are merely illustrative. They do not reflect actual conditions, although they could point to some genuine dangers. Political implications and economic costs of food dependence serve to highlight the gulf between poor countries and wealthy ones. That cannot be the only factor taken into account in order to accurately reflect the food situation. Disparity is not only apparent from one country to the next, but it is even more apparent within specific human groups. Previously, world hunger was conceptualized as a problem of geographic regions. Today, world hunger is being revisited as an issue of class. As far as food is concerned, poor peasants in Nigeria or in Mexico, both oil-exporting countries, probably have more in common with each other or in common with poor peasants in Brazil, which imports fuel, than with wealthy urban classes in their own countries. Those poor peasants do not receive imported food, subsidies, or aid. They cope by relying on their own resources. Those resources are additionally compromised by the inevitably high price of exploitation paid to their dominators. Their diets resemble diets of half a century ago more than they resemble diets implied by the expansive behavior of the world market in which less than 15 percent of world production shares. The total cereal deficit of poor countries is just a little more than 8 percent of their own production. Globally, it is not the lack of food that produces hunger and malnutrition. World production, without being excessive, would be enough to adequately feed everyone. Hunger starts with unequal distribution.

If dependence and hunger are not equivalent, they are related. As dependence becomes more acute, hunger almost inevitably spreads. Much data confirm that this is exactly what is happening. It is also clear that hunger and malnutrition, the harshest expressions of disparity, are unlikely to be effectively confronted without regaining self-sufficiency and dietary self determination: the ability to freely select suppliers, the type of goods deemed necessary, and target objectives. This sphere, the recovery of dietary sovereignty, necessarily involves the nation, the state, and the population at large. It includes peasant producers as the central subjects of such a project. The illusion that the world peasantry would disappear with modernization projects, with the industrial millennium, has been shattered. The myth of a waning peasantry has been dispelled more effectively in the real world, where the facts speak for themselves, than in ideology, where such a delusion persists and still influences many of the projects and the actions of those who govern.

The industrial millennium did not fail. On the contrary, it has come and gone. We are on the threshold of a new era in manufacturing production that still lacks a suitable name. Any promises the industrial age held out have already been kept. Now we know for certain that those promises were limited and narrow. Many of the assumptions of the industrial millennium were nothing more than that. It was a mistake to assume that Third World countries simply would follow the same compulsory and universal pattern of growth as the industrially advanced countries. Progress and wealth that overflow from above, eventually trickling down to those below, never materialized. No homogeneous universal culture arose: enlightened, rational, Western. Instead, a new dimension of inequality emerged. A new dimension of human diversity as an enduring condition emerged as well. In that very diversity resides hope.

The Syndrome of Inequality

INVENTING THE FUTURE

The long history of corn would not be complete without an account of its modern expansion and growth. This never was more rapid than in the 1980s. Between the late 1950s and the early 1980s, world corn production tripled, but the total surface area devoted to corn cultivation worldwide grew by less than half. This implied that corn yields slightly more than doubled. Such a remarkable accomplishment meant that production was able to keep a step ahead of incredibly rapid population growth. Corn had never been more abundant. Many other foodstuffs experienced a similar trend, albeit at lower rates. Production of cereals worldwide grew 134 percent over this period. Like corn, the rate of cereal production also surpassed the rate of population growth, although by a slimmer margin. Despite these strides, malnutrition and hunger did not disappear. They most likely grew and affected an even greater part of the world's population in the 1980s than ever before. Dietary disparity became so severe that even these increasingly higher cereal yields merely served to check hunger rather than do away with it. Production would have to grow faster still in order to eradicate hunger altogether. It makes sense to ask ourselves, briefly and simply, how such growth might come about, its agents and forms, its costs and limits. I will try to respond to such queries as far as corn is concerned. I do not contrive to predict the future. Rather, I attempt to contribute to a hypothetical construction of what we might expect. I do not try to prophesy about what lies ahead. I do suggest another way to analyze the past and the present for the sake of inventing the future.

Corn is produced many ways worldwide. Such diversity is one aspect of corn's many virtues and one of the reasons for the ready adoption of corn. This diversity can be better understood for purposes of analysis by sorting the many ways corn is produced into smaller groups defined by some common characteristics. The modes constructed with that purpose in mind are not descriptive; they do not reflect any concrete empirical reality. Rather, I have separated out some of their abstract traits in order to better understand their behavior. With these predictable precautions in mind, I divide corn production into two modes. One mode is represented by capitalized intensive agriculture, also known as scientific agriculture or production by the wealthy. The other mode is represented by traditional peasant agriculture. This mode can be extensive or intensive in its use of human labor and utilizes relatively few resources beyond those readily available and controlled by the productive unit. This is farming by the poor.

Probably more than two-thirds of world corn production takes place under the conditions and norms of scientific or capitalized intensive agriculture. A little more than 40 percent of the world's corn is produced in the United States. Capitalized intensive agriculture was responsible for and remains the force behind most of the growth in corn production worldwide. Between 1961 and 1965 and between 1981 and 1983, developed countries added an average of 121 million tons a year to world corn production, as opposed to the 44 million tons produced by underdeveloped countries or the 55 million tons produced by the former Soviet bloc countries. The leading indicator of growth in capitalized intensive agriculture was the increase in yield per acre. In the United States between 1950 and 1981–83, corn yields per acre rose from just under a ton to very nearly three tons, an increase of 191 percent. The surface area devoted to corn cultivation fell by 22 percent, while the volume of corn harvested increased by 125 percent. As producers in other developed countries and in some privileged sectors in underdeveloped countries began to adopt U.S. technological innovations, similar tendencies appeared there as well. Between 1961 and 1965 and between 1981 and 1983, developed countries only brought an additional 0.13 acres of corn under cultivation for every additional ton of corn they harvested. Developing countries, on the other hand, had to bring an additional three-quarters of an acre under cultivation for every additional ton of corn they produced.

There is no question that scientific agriculture has been tremendously efficient in the past. This has been especially true for corn cultivation. Now the pertinent question revolves around what we can expect of scientific agriculture in the future. There is no easy answer. The stagnation of scientific agriculture has been predicted more than once, and has come to nothing. It is also true, however, that it has been necessary for developed countries and poor countries alike to prop up scientific agriculture. The governments of wealthy nations provide direct assistance in the form of large subsidies. Poor countries are forced to make enormous payments for foodstuffs on the world market. Even if those conditions prevail, it is not at all clear if growth at a similarly fast pace can be expected to last into the future.

The main component of growth in capitalized intensive agriculture, the increase in yield per acre, has an open technical frontier. In corn trials conducted in the United States on small test plots, techniques mimic fields overly invested in labor. Corn yields of nearly nine tons per acre, a figure not obtained for any other cereal, are common in these test plots and are three times higher than results obtained in practice. Prime land is selected for test plots, which can only be imitated on a larger scale to a very limited degree. Even so, if that type of care were extended gradually to all the farms or enterprises using existing techniques, a steady growth in corn yields would be the result.

Obstacles to such a scenario are basically economic. The small parcels used in test plots are worked almost manually and each plant receives individual attention. In developed countries there is a relative scarcity of labor. Labor costs, therefore, constitute the most expensive factor of production. Raising productivity by making agriculture even more labor-intensive is not economically viable. Strides have been made toward improving the efficiency and accuracy of agricultural machinery in carrying out uniform tasks at higher speeds, but machines have yet to be designed with the flexibility and discretion of human labor. Some experimental machinery, robots, has been constructed to carry out such simple tasks as picking fruit. Such tasks, nonetheless, require complex identification and derived decision making. The functioning of such robots is still awkward and imperfect and their cost too high. There is a certain consensus that agricultural machinery, while becoming increasingly fuel efficient, will continue to improve, but functioning essentially as it does now: treating land and crops as though they are equivalent. Mechanization will be able to improve the quality of the tasks it is to accomplish to some degree, and perhaps even carry out these tasks at higher rates of speed and at lower costs. Mechanization does not seem to be the most appropriate vehicle to reconcile productive potential and available technology. However, mechanization can be a weighty factor in determining the limits of increasingly larger territorial production units.

The rise in corn yields since the 1930s owes much to genetic engineering. The frontier for advances in that field is still wide open and can be expanded even further with new knowledge and new techniques. It is possible that smaller plants may be developed, more of which would fit into the same surface area. This would make more efficient use of solar energy. Plants with greater resistance to known pests might be developed. By the same token, these same plants might be more vulnerable to yet other pests by virtue of their genetic uniformity. The incidence of corn smut in the United States in 1971 was very high precisely because of the genetic uniformity of plant varieties. Although it is uncertain and far off, perhaps perennial as opposed to annual corn varieties will be created from teosinte, native to Jalisco, Mexico.

Such a development would imply an enormous reduction in costs. Almost anything seems possible in the field of genetic engineering, although it will take time. The qualifier "almost" is crucial in this case. Genetic selection and genetic engineering modify individuals and groups of living beings, but they do not alter certain aspects of life itself. There will never be plants that grow without light, without water, or without nutrients. There will be advances in genetic engineering, but there never will be miracles.

The intensive use of fertilizer is part and parcel of such improved seed. New varieties will require more soil nutrients, more fertilizers, in order to continue to produce higher yields. Herein lie the greatest economic and technical restrictions. There is an inverse relationship between amounts of fertilizer and rising yields. When applied for the first time on land that has not been fertilized previously, fertilizers have a tremendous impact on yield. The effect on yields is much greater in these fields than it is in fields already subject to the intensive use of chemical fertilizers. Such fields already have relatively high yields as a result of the application of fertilizers. The experience in the U.S. corn belt illustrates this tendency. When fertilizer was applied for the first time, each pound of fertilizer produced twenty-four additional pounds of grain. With a second application, each additional pound of fertilizer only produced an additional twelve and a half pounds of product. Upon the third application, an additional pound of fertilizer produced only eight additional pounds of grain. By the fourth application, just over three and a half pounds of additional product resulted. Finally, by the fifth application, each additional pound of fertilizer yielded not even an additional nine-tenths of a pound of grain. For that reason, in order to elevate yield to 150 bushels per acre, a little over four tons per acre, over and above the 110 that are obtained today, the application of nitrogen-rich fertilizers would have to increase 400 percent. In any case, in the most developed countries today the rate of fertilizer application is already at or very near the point of diminishing returns. For this reason, any increase would suppose a negative return. This technical relation is translated into a cost-benefit economic paradox. If the cost of fertilizer were less than the grain, there would be no upper limit in the application of fertilizer. That, however, is not the case. On the contrary, the industrial production or manufacture of nitrogen is an expensive process, with very high consumption of fossil fuels. Furthermore, the production of fertilizers is very concentrated and subject to scarcity, factors working against significantly lower prices as a natural, market-driven tendency.

Available technology does not offer alternatives that might significantly alter the relationship between yields and the application of fertilizers. Some operational measures, such as crop rotation, no-till, the use of herbicides, and so forth are part of today's technical repertoire. These can lessen costs or increase yields marginally, but only will barely be able to counteract the negative return fertilizers ultimately bode. This limitation is an expression of the low energy efficiency of capitalized intensive agriculture. Poor energy efficiency also is reflected in environmental risks, in the destruction of resources in order to increase earnings. Technology has done little to resolve this basic contradiction in scientific agriculture. Agriculture is deemed to be scientific because it understands and dominates some variables of the complex process of the growing cycle. However, scientific agriculture has yet to focus on other variables in that process that are proving to be its most challenging limits in the present.

The growth of capitalized intensive agriculture by bringing new lands under cultivation cannot be dismissed as a possibility. It does have severe limitations. In some developed countries, full agricultural capacity has practically been reached, as in Western Europe. Even so, it is possible that some crops like corn can gain surface area by virtue of their high yields. They will do so, however, at the expense of other crops. The net gain only will be that of the productive differential between the two crops. In the United States and Australia, there is an agricultural frontier of land that either is of lesser quality or that requires costly investment. The elimination of the crop land reserve program in the United States since the crisis of 1972 did away with land reserves as readily available short-term resources. Climatically vulnerable and poor lands remain, but it is highly unlikely that these will afford yields similar to those garnered today.

The growth of capitalized intensive agriculture in developed countries is possible. It is unlikely that growth could sustain the rates and levels that it has enjoyed since World War II. For many crops, such as wheat, the increase in yields slowed dramatically during the 1970s. This is not the case with corn yields, an exception that goes a long way toward explaining corn's growing importance. It has not been possible to substantially raise the yields of other crops, such as soybeans, that respond only in a very limited way to fertilizers. Technical factors restrict elasticity. These do not imply, however, that growth is impossible.

Political and economic decisions of governments in developed countries can have an appreciable effect on the extent of growth in the production of foodstuffs there using scientific agriculture. After the crisis of 1972, for example, the United States made the decision to export grain instead of fertilizer. This modified the relation between the price of fertilizer and the price of grain. It also affected subsidies to U.S. farmers. A subsidy for fertilizer is not out of the realm of possibility, but it would have to be on an enormous scale in order to compensate for decreasing yields. If Western Europe or Japan were to eliminate their protectionist tariffs, their domestic production would most likely fall and their imports most likely would rise. If the price of grain were to increase in a substantial and accelerated way, at the very least the market for grain in the Third World would come to a standstill, and probably decline. There are other economic obstacles that might speed up or delay the growth of the production of foodstuffs in developed countries. Such obstacles are as harsh or harsher than any technical obstacles affecting agricultural earnings. The comparative advantages for wealthy countries as producers of foodstuffs are not absolute or sacred. They are relative and variable as a function of the economic policies of the states that create and maintain the margins of return for capital invested in agricultural production. Here there is little hope for fundamental change. As has been demonstrated in the past, wealthy states tend to act in their own strategic and economic self-interests. Decisions of that nature probably will be repeated in the future and will affect technical conditions as much or more than scientific discoveries. Political decisions appear today, perhaps more than ever, as decisive factors in scientific agricultural production.

Even if the production of foodstuffs were to grow at an accelerated rate in wealthy countries, that would not guarantee any improvement of general nutritional levels or eradicate malnutrition. The growth of the world market for food has not had that effect in the past. The largest part of the foodstuffs that are traded on the world market are destined to change diets, not improve them. The expansion of meat-based diets always implies an increased cost for a market basket of food, the rise in the consumption of agricultural products per capita. It does not imply better nutrition. Neither does the substitution of one cereal for another as the leading staple food have an overall beneficial impact on nutrition. Such is the case when the participation of poor countries in the world market is limited to supplying wheat to the relatively privileged urban sectors of the population of those countries.

The increasing supply of foodstuffs provided by developed countries has fueled the tremendous growth of the world market. This has not corrected the poor distribution of foodstuffs but, rather, has made it worse. In many cases, pricing mechanisms have been responsible for inhibiting or pushing aside locally produced cereals or other foodstuffs. Growing dietary dependence on the part of poor countries does not coincide with a general improvement in nutrition, although it may have profound repercussions for the transformation of the diets of the best fed sectors. Any further increase in agricultural surpluses in wealthy countries would simply spur the perverse growth of a world market that has not contributed to checking world hunger.

Once the international community became aware of this situation, there was a series of multilateral efforts to increase the production of food in the Third World. One way to do this that seemed logical and obvious at the time consisted of directly transferring techniques associated with high yield scientific agriculture to poor countries. If previous experience were any indicator, the results of such a strategy would be guaranteed. Thus the Green Revolution was born. It consisted of providing poor countries with rigid technological packages loaded with those factors responsible for the high agricultural yields in wealthy countries: genetically engineered seed, chemical fertilizers, mechanization, and so forth. Those packages demanded concentrated, intense capital investment. They were introduced in areas where very low capital investment in agricultural production prevailed. Many countries could not even contemplate the adoption of such a model, in spite of the available international aid, because it was simply too expensive.

Countries that chose scientific agriculture were forced to overinvest their meager resources in delimited areas and privileged sectors. In this way, enclaves of capitalized intensive agriculture were created in countries with otherwise insufficient aggregate agricultural investment. Agricultural development in these countries amounted to showcases that monopolized public and private resources. When it was successful, as in the case of wheat cultivation in Mexico, the initiative could not be widely applied. Continued success required mounting subsidies. It is worth remembering that the initial nucleus was a group of producers favored with good lands, irrigation, and generous governmental support: a privileged group compared to the majority of peasant agricultural producers. Technical change generated profit levels that were similar to or greater than those of farmers in the wealthy countries. Those profits, however, were not reinvested in agriculture. They were transferred to safer and even more profitable investments. Areas enlisted in the Green Revolution became a bottomless pit of government subsidies. Often, the cost of the commodities that were a product of these pet projects was higher than the cost of importing those same goods from abroad. Poor agrarian countries were introduced to senseless, absurd extravagance. The contradictions that arose when the agriculture of the wealthy came face to face with the agriculture of the poor in the production of foodstuffs led to domestic problems. The exportation of Mexican wheat at huge economic losses for the government did not mean that the demand for food in Mexico had been met. It only meant that the prevailing condition was one of high capital investment in wheat, despite growing inequalities and structural inflexibilities that prevented the domestic use of locally produced foodstuffs.

Not everything about the Green Revolution was counterproductive or trivial. It was a costly learning experience. It led to research on genetic engineering in agricultural climates other than those of temperate regions. Genetically improved seed finally was separated from the rigid aid packages for which they had been created. Peasants and grass-roots organizations led initiatives to appropriate that seed. Such actions offered important perspectives for future development. They especially demonstrated that scientific research had an important task with respect to the agriculture of the poor. Many psychological and institutional obstacles had to be surmounted in order for such a task to acquire the status and direction merited by that challenge. The main objective for research continued to be an increase in yields, a norm derived from other conditions. No attention was given to other alternatives that could raise aggregate territorial production. These included the lessening of uncertainty, bringing new lands under cultivation, and the elimination or reduction of fallow. Many of these already had stood the test of time, but they were not so well known as those that had been part of a more conventional road to capitalized intensive agriculture.

The other modality, agriculture of the poor, is more complex and diverse. Agriculture of the poor is a heterogeneous ensemble. It lacks a unifying norm such as profit. Or more accurately, what defines this mode is much more elusive and complex. Perhaps it can be expressed as some sort of vague concept of cultural, social, and physical survival. Although vague, such a concept is far from static. It changes over time and according to circumstance and always implicitly carries with it the desire to improve, the equivalent of progress, but understood as escaping a condition of subordination, of exploitation, of structural restriction. In spite of that, and perhaps because of it, the agriculture of the poor never has remained motionless.

If we were to use the behavior of aggregate domestic production as an indicator, hardly an impartial criterion but the only one available to us, then the production of cereals in Third World countries grew in the same proportion as that of the wealthy countries. Poor countries even outpaced wealthy countries in the production of some commodities, such as wheat or rice (not the case with corn). Not only that, export agriculture also grew in poor countries, and much more rapidly than the production of foodstuffs alone. Despite this, populations grew much faster in Third World countries and in some cases rendered any increases in agricultural production inadequate.

Inventing the Future

Agricultural growth was uneven among the heterogeneous countries typically referred to as developing nations. In many of them, that growth was less than average and did not meet the needs of individual countries, which resulted in even greater dependency. In any case, peasants dominated Third World food supplies. Peasant production grew at the same rate as that of wealthy countries, and in many cases was pushed to the brink of collapse.

It is much more difficult to pin down the most important factors in the growth of the agriculture of the poor. More strictly speaking, it is difficult to generalize some of the known factors. Large-scale statistical constructions omit or inadequately reflect the fluid conditions of peasant agriculture or reflect them in a distorted way. Peasant agriculture is made up of factors that are triggered by social networks of exchange rather than by marketing or accounting mechanisms. Aggregate figures are inadequate to alert us to such factors of production that cannot be explained using conventional capitalist categories. The statistically abnormal is often the very thing that is usual, reasonable, and historical. It can be measured, and it would be very useful to do so. But we are not doing it. Conventional statistical categories do not lead us in that direction. They have a conceptual and ideological basis that hampers or ignores it. They are derived from a uniform and narrow understanding of progress that aspires to measure something that has not yet come to pass.

The importance of an increase of harvested acreage as a factor in rising production levels in the agriculture of the poor is greater than what statistics might suggest. According to available figures, between 1961 and 1965 and between 1981 and 1983 cultivated acreage in corn in developing countries grew by 30 percent, acreage in wheat by 36 percent, and acreage in rice by 14 percent. Land dedicated to the cultivation of foodstuffs increased by almost a third. Virgin lands were added only exceptionally. More often marginal lands were brought under cultivation, lands low in fertility or climatically vulnerable that had been used for other purposes, such as pasturage or gleaning. The yields on those lands were almost always less and their harvests subject to greater uncertainty. In not a few cases, the growth in cultivated surface owed to the more frequent sowing of lands that previously had required a period of fallow according to traditional extensive techniques. In shifting cultivation typical of the humid tropics, the reduction of fallow resulted in lower yields and a reduced number of species used in mixed farming. The elimination of fallow was explained in other cases by the addition of chemical fertilizers, as happened in the Mexican plateaus, which led to more frequent cultivation with no corresponding rise in yield per unit. The use of fertilizers in many parts of the Third World was not presented as an option to raise yields, but as an unavoidable necessity in order to keep production levels steady. The lack of research on this issue probably was responsible for a large loss of nutrients, although it also was likely that other research, the accrued peasant experience, indeed had taken it into account.

Bringing marginal lands under cultivation does raise the productivity of the land. It also reduces the productivity of labor necessary to cultivate it and implies low yields and substantial risk, frequently courting disaster. Cultivation of marginal land also entails a loss of resources and of means. In many parts of the Third World, the cultivation of land that was formerly pasture for draft animals has forced the adoption of mechanical traction and additional factors of production that must be acquired in the marketplace at high prices. Autonomy on the part of peasant producers has been lost. With that loss important dietary alternatives are closed off.

As peasants continue to increase in numbers and needs, they put marginal lands under cultivation because they do not have access to other, better lands. Considered as a whole, Third World countries have uneven agrarian structures characterized by a high concentration of the most valuable territorial resources in a few hands. That high concentration is exacerbated by an irrational use of the best lands, another reflection of the inequities that tend to fragment Third World societies. Often, the best lands are reserved for export crops. Excesses are not uncommon. In many places sugar cane is irrigated even where other suitable land is available to sow sugar cane seasonally. Irrigation guarantees a crop with high yields and higher profits for sugar cane barons. Neither is it unusual for landowners to devote areas suitable for highly productive cereal agriculture to the extensive raising of livestock. Livestock is more profitable than cereals due to strict price controls, prices linked to those of the world market, which affect the price of staple foods but not those of luxury foods. Inequality and irrationality in land use are constituent parts of the agrarian structure of developing countries.

Bringing new, suitable lands under cultivation continues to be the fastest and cheapest means to increase agricultural production. When the increase in the cultivated surface is carried out by peasant units, there are immediate and direct effects on the diet of the producers, those most affected by poor nutrition. It also has aggregate effects on the systems of supply and on the domestic economy through a better distribution of income. The expansion of cultivated surface area is one effective, although perhaps ultimately inadequate, way to confront problems of nourishment and to increase agricultural production in poor countries. The means to do so, independent of the natural attributes of territorial resources in different countries, is dependent on agrarian reform, on the redistribution of access to land. Such a statement is not speculative. It has been borne out by repeated historical experience. Agrarian reform as an economic strategy is premised on a political decision that is on the horizon of almost all poor countries the world over.

Agrarian reform is not a trendy topic today. There are even arguments against it, given recent bad experiences with marginal and limited agrarian reforms. Those agrarian reforms typically came down from above and at times from outside, as in Latin America and reforms stemming from the U.S.-sponsored Alliance for Progress in the 1960s. These coincided with the expansion of international food aid. The topic of agrarian reform endures, despite fashionable trends to the contrary. The structure of landownership and access to land no longer respond to the characteristics and necessities of the poor countries. Rather, they belie it. Perhaps for that reason, prior experiences with agrarian reform do not constitute a suitable model for the present and the future. The simplistic dichotomy between private property and state property is not an adequate framework within which to realign agrarian structure with social structure in poor nations. Other proposals are brewing to resolve the preeminence property systems exercise over society at large. Agrarian reform is acquiring new substance.

Better yields also were a component in the growth of agricultural production of poor countries. From the late 1960s to the late 1980s, yields in developing capitalist countries grew substantially and at rates just slightly lower than those of wealthy countries: cereal production grew by 45 percent, wheat by 64 percent, corn by 38 percent, and rice by 41 percent. Over the same period, developed capitalist countries saw cereal production rise by 46 percent, wheat by 44 percent, corn by 66 percent, and rice by 13 percent. A significant part of rising yields could be explained by an agricultural dichotomy in poor countries. The biggest gains in yields owed to the capitalized and modernized sectors within Third World countries. Mexico is a perfect example. Mexican wheat yields were more than double averages worldwide and one of the Green Revolution's great success stories. But peasant producers had very little involvement with wheat production. Wheat cultivation was confined to irrigated areas and was dominated by modern capitalist agriculture barons. With many variations on a theme, the Mexican example was repeated all over the Third World.

Yields in peasant agriculture in poor countries grew in spite of any skewing effect of heavily capitalized agricultural enclaves. Peasant agriculture grew more slowly than capitalized sectors within those same countries, although the full extent of that growth could not be precisely determined. In many cases, growth in peasant agriculture owed to a greater intensity of labor to ensure the best possible development of the plants. Occasionally, this was at the expense of the productivity of labor, as measured by input-output. Often, that additional labor input was the logical outcome of an absolute lack of more productive alternative activities within a particular social context. Such a lack of alternatives led peasants to see what scientific agriculture would call the overinvestment of labor as simply more work, and greater employment as a means to a more plentiful harvest. There was an abundant workforce in almost the entire rural underdeveloped world. That labor could not always be incorporated into agricultural tasks due to seasonal restrictions. When planting only lasted ten days, a corresponding level of labor intensity for the rest of the agricultural cycle was up against very challenging limits. Usually, it was impossible to exhaust all available labor on the small area accessible to peasant producers. The misalignment of a readily available workforce and obstacles impeding their productive employment had not been a topic of agronomic research, among other potential subjects that merit investigation. When certain plant varieties thrive during the agricultural calendar, whether they are resistant to drought or frost, whether different varieties and species could be combined in the same field and in the same cycle, answers to all these questions could favor fuller employment and increase yields just as much as raising the productivity of one lone variety. Uniformity corresponds to capitalized intensive agriculture. Diversity, which at various times has been misinterpreted as disorder and chaos, corresponds to peasant agriculture.

Rising yields in peasant agriculture owed to many factors. Some yields responded positively to genetic engineering, to the adoption of more productive varieties. In some instances, genetic improvement was the result of experimentation and selection made by the peasants at their own initiative. In others, genetic advances such as miracle rice or the wheat developed in Mexico by international research teams were the products of scientific research that trickled down to peasant producers by a number of different means. Yields also improved due to the adoption of chemical fertilizers and insecticides, occasionally applied in excess and at serious risk. Scientific agriculture also developed some resources that directly addressed peasant needs. The possibilities of development in this vein were wide-ranging and have scarcely scratched the surface. It also was the case that the rising yields were sometimes nothing more than an illusion, derived from the very things that managed to elude the statistics. A case in point was when, due to falling fertility or the adverse application of some chemical, mixed farming was abandoned in favor of planting one sole species. This raised a specific crop yield, but lowered the yield of the land overall and lowered the availability and variety of other foods.

A myth persists that peasants resist adopting scientific advances. Many believe such a myth. Such resistance does not exist. What frequently does exist is a misalignment that appears when peasants are subjected to the generalizations of scientific recommendations, recommendations converted into dogma but not backed up by research on or knowledge about the needs of peasant producers. Highly productive seeds that also are highly demanding of investments that the typical peasant cannot make or that are highly vulnerable to known risks are naturally rejected. Peasants also refuse to apply fertilizers, for their cost outweighs any additional production. Peasants decline to adopt expensive machinery that rather than making work easier only makes it redundant. Those and other examples, repeated thousands if not millions of times in the Third World, are used to bolster the myth of the impervious peasant. They also are at the origin of a well-founded distrust on the part of the peasant with respect to scientific recommendations, especially when the costs of any error are paid not by the person who makes the recommendation but by the person who follows it. The truth is that scientific agriculture still does not have much to offer peasant agriculture. There is no interest in it. Its recommendations, almost always derived from capitalized intensive agriculture, are not necessarily false. They simply are far removed and inadequate when applied to other ways of producing. Scientific research was initiated, after all, in order to come up with recommendations and means useful for developed countries. That simply has not been done in poor nations. Agronomic research, for and in collaboration with peasants, is an encouraging and important process, but basically has yet to begin in earnest.

Increasing yields is possible and absolutely crucial to the agriculture of the poor. It is the fastest and easiest means to deal with hunger and to raise agricultural production. Corn, for example, has an average yield of more than three-fifths of a ton per acre in developing countries, less than one-fourth the U.S. average. Wheat production is nearly two-thirds that of wealthy countries, and rice less than half. The trick to achieving higher yields is no secret. It is known and it is accessible. Yet agriculture in poor nations is very far from the point of saturation as far as the productivity of fertilizers is concerned. The application of fertilizers in poor countries would have a much more dramatic effect than it could have in wealthy nations. To raise yields in corn production by a ton per acre in poor countries is theoretically easier and more economical that raising it by the same proportion in areas where capitalized intensive agriculture already exists. Even after an increase of that magnitude, corn yields among poor nations would be scarcely a third of the average yield in the United States. If corn production in existing plots in developing countries would have risen by even half a ton an acre in the 1980s, a modest target, production worldwide would have risen by 66 million tons. Increasing corn yields by less than half a ton an acre in the wealthy countries, an ambitious target, would have resulted in an increase worldwide of just over 44 million tons.

That line of reasoning had a lot of logic and truth to it. It was put into action with a vengeance after the crisis of 1972, when many of the programs promoting integrated rural development were founded. Since that time, the ambitious targets of wealthy nations were more easily met than the more modest and realistic targets of poor nations. From 1974 to 1976, U.S. corn yields grew by over three-fifths of a ton per acre and yields in developing countries grew by less than one-fifth of a ton per acre. Wheat yields grew by over a fifth of a ton per acre in wealthy countries and by just over one-tenth of a ton per acre in poor countries. Rice was the only grain for which the relation was reversed. This experience did not dispute the viability, the logic, or the urgency of the objective. It did reveal unforeseen obstacles. Among these obstacles was a paradox worth mentioning. Long-term programs to raise yields in poor nations that followed the model of developed nations were very expensive. It made a lot of sense to pay those costs when the price of grain on the world market was high and rising. But when these prices went down, almost always as an expression of agricultural subsidies in wealthy countries, the economic viability of raising yields in poor nations fell and many programs were discontinued. It was clear that, in the long run, the political and economic cost of dietary dependence would be higher than dietary independence, but few Third World governments were thinking about and planning for the long term. That planning was a necessity that could not be put off, but it also was a luxury that could not be paid for. To live in the present and be unable to plan for the future is exactly one of the hallmarks of poverty.

Other obstacles refer to the lack of applied and basic research into how to adapt techniques that are known to be effective and into how to find alternatives to those techniques that are effective but deemed unsuitable or too expensive to be widely applied. Raising productivity through the increased use of fertilizers is above reproach when fertilizers are sufficiently available and applied at the most opportune and optimum rates. Thus fertilizers cannot be applied effectively when reliable meteorological series, knowledge of the soils, or familiarity with the extraction of soil nutrients are not available. In any case, research capability is almost nil and standing research is often unsuitable. It is far removed from peasant existence and does not offer peasants solutions. Research must be deeply rooted in peasant production. Aid must be straightforward and direct. This aid must be in the form of complementary services, such as credit, provision of factors of production, and trade programs with the goal of increasing the autonomy of producers, and not have the unintended effect of promoting their further subordination. Only measures such as these can ensure that economically feasible, viable, and unequivocal means of raising yields in peasant agriculture become a reality. The sustained political will to do so, as in dedicating an important quantity of public resources to agricultural productivity, is likewise a requirement for such an alternative.

These two paths, agrarian reform and integrated rural development, have become separate paths, alternative routes. In general, governments have opted for the second with the hope that it would permit them to avoid the agrarian problem altogether, to buy time until the much-anticipated extinction of the peasant came about. This has not happened. On the other hand, there are those who argue that the simple surrender of land solves nothing. That is so when only an insignificant fraction of land is in play. But when territory is redistributed to the point of radically altering the agrarian structure, then the simple surrender of land does not exist. Neither does the simple surrender of the vote exist when the majority does not have it. There is no simple process when power is being redistributed in society. The redistribution of power is the real issue at hand when we speak of eradicating hunger and of regaining dietary self-sufficiency. The redistribution of power is implicit in all the paths, paths that in the end will all converge without exception.

The only way to confront the problem of world hunger is to increase peasant production using the many and at time unimaginable means to achieve that goal. To think in these terms threatens myths and truths that have been proclaimed absolute. To accept a future in partnership with the peasants of the world is not something that inspires confidence among those who worship progress or those who hold the reins of power. It no longer is a problem of beliefs, but one of evidence that we must acknowledge. It is obvious that the peasants of today in no way resemble those of yesterday. Nor do those peasants of the past resemble those who came before them. Even if we call today's new peasants by another name, the basic issues remain the same. Hunger is one of those issues. A future without hunger is possible in the not-too-distant future. It is difficult to achieve, but it is not too much to ask.

: 15 :

BRIEF REFLECTIONS ON UTOPIA
& THE NEW MILLENNIUM

Now this book draws to a close. In the social sciences this is customarily the moment of truth, the final opportunity to summarize the research in the preceding chapters by laying out an ordered series of simplistic and positivistic conclusions. The only thing I am sure of is that I do not have much more to say. I do not have a general formula, or some sort of universal prescription or law to put forward. This book is a product of my profound dissatisfaction with what are generally accepted as self-evident truths, with deductions derived from universal models. I am critical of versions of history that are nothing more than a series of morality plays or exercises in self-congratulatory rhetoric. I cannot accept history as rationalization, history as a series of foregone conclusions proclaimed from a position of triumph and power. To a certain extent, this work is a critique of more conventional histories that use academic exorcism as a substitute for theory. These conventional accounts typically act as a means to justify and classify historical processes rather than explain them.

That is not to say that we must wipe the slate clean and start from scratch. I take full responsibility for the theoretical position that I have assumed, am perhaps openly fond of, and even overtly biased toward. I offer my interpretations and my conclusions from a perspective that is my own. It seems to me that this book is full of such explanations and findings. Each chapter has one, several, or many of what I consider to be conclusions. This could be the time to beg forgiveness if, as I fear, I have committed the deadly sin of interpretation, of confusing density of information with theoretical depth. Terror of committing such a sin precludes any attempt to summarize my conclusions in this final chapter and present them as my worldview. I do have one conclusion of that caliber, but I suspect that it is neither original nor interesting, or even pertinent here. This research does not attempt to explain the world. It attempts to analyze some social processes in which corn has played an important role. That analysis supports the conclusions I have included in this book.

It remains to be seen if the cluster of processes, as a whole, has something interesting to offer.

I believe that they do. I beg forgiveness for my arrogance, which is the prerogative of the author. Again, many of the conclusions in this wider-ranging dimension have already been stated in the preceding chapters or are implied by virtue of contrast and counterpoint between the different processes. If as a writer I have been effective, everything has already been said. I have my doubts in this respect, which is why I am going to reiterate something that I already said or that I believe I should have said if I did not. While my findings are not necessarily original or novel, I will not risk the possibility that they go unheard. They deal, essentially, with the issue of imagining the future.

I have been able to document diverse processes in which corn has had an important role: as a conqueror of wide-open spaces in China or the United States; of the seasonal agricultural calendar in southern Europe; of the organization and intensity of labor in Africa; as sustenance for workers in profitable, dynamic, and new enterprises; as a way to stave off hunger in the past and, I hope, in the future; and, lastly, as a symptom and stigma of poverty. In short, I have been able to document corn's role as one of the chief renewable resources in the founding of the modern world. Corn is not the only American plant resource that plays such a role. My penchant for corn does not go that far. I simply study the data and am convinced that they speak for themselves. Whatever may have been the course corn followed in its migration abroad and whoever may have introduced and promoted the plant, the role of this American cereal becomes more important in times of accelerated transformation, in circumstances of rupture and disjuncture. Under those conditions, corn's botanical virtues are essential for its timely adaptation under powerful and sometimes even intolerable pressures.

The most violent disruptive conditions in the processes analyzed in this book are those created by the expansion of Western capitalism understood in the broadest sense of the term. That sense does not only refer to capitalism as a mode of production, although it includes it. It also includes the forces unleashed by capitalism that went to make up a global system of direct or indirect political and economic domination. It refers to the ideological, technical, and cultural apparatus, to those things that in a certain sense could be grouped together under the concept of a great cultural trunk or civilization. There is then a clear link between the expansion of corn and the expansion of world capitalism, often simply referred to as modernization. It would be accurate to say that corn is a central character in the history of capitalism, but it would be unjust to suggest that the migration and spread of corn is its only conceivable circumstance; resistance to the encroachment of corn is another. Corn is also so much more than that. It is a unique resource for the construction of a new reality, for change and social transformation.

Capitalism traditionally has been interpreted and touted as a more advanced level of development. Such a paradigm either surreptitiously or overtly permeates the widest-ranging social theories. Implicit in this paradigm are judgment calls, value judgments, and the absolute faith in the universal maxim that all of humanity will pass through capitalism, the most advanced state of human development. Faith in such a chain of events transformed capitalism into a millennialist prophecy. That prophecy did not have its origins in some divine design, but in the inevitable laws of history. Western capitalism was and still is a promise of global homogeneity. Western-style capitalism was to be the driving force behind unity and, in its most generous assessment, of equality. Capitalism, often spoken of as modernization and economic development, was promoted as the wave of the future. It would hasten the demise of all previous precapitalist forms, obsolete and condemned, and bring about their total eradication. These premises and suppositions are stated with more subtlety and diversity today, but they continue to dominate social thought, ideology, and political agendas with maintaining or achieving hegemony as their goal.

The processes analyzed in this work are not sufficient to allow for general speculation about such suppositions. They are significant enough to subject many of the suppositions typically associated with capitalism to criticism. The first supposition I would like to challenge refers to homogeneity, to the uniformity that capitalism will impose on production, consumption, thought, and behavior, on culture itself. The tendency toward uniformity exists and is one of the consequences of the expansion of capitalism, although it frequently is overstated and overlooks the persistence and the emergence of fundamental differences. That effect is limited and not general. The tendency toward uniformity affects social relations and specific social classes or groups with different degrees of intensity. In any case, the impact of the tendency toward uniformity is much less than the tendency toward diversity promoted by the spread of capitalism. The tendency toward diversity is the more obvious and important of the two, the dominant tendency.

The processes analyzed in the previous chapters illustrated the heterogeneity and diversity with which different societies and human groups utilized corn and the processes of change generated by the expansion of capitalism. Those processes even promoted an increase in biological diversity. Corn's migration produced an increase in the number of its varieties and races worldwide, derived from the crossing or free pollination of plants of different origins and traits. The opposite situation, that of the United States, was presented as an exception to this general rule. In the case of that country, the wide distribution of hybrid seeds produced from a limited number of stable varieties demanded that measures be taken to preserve what formerly had constituted an enormous genetic diversity. Overall, the increase in diversity was more widespread and common than the cases in which a process of reduced botanical diversity took place.

The same thing occurs with the techniques and methods for corn cultivation, already quite diverse since before corn's migration abroad, when those techniques and methods multiply in order to adapt to new environmental and socioeconomic conditions. The variety of ways to grow corn today is much greater than in the past. Many of the old systems have not disappeared, although they have been transformed, because those systems have not been able to be prevail over their own circumstances. None of the innovations, from the use of draft animals to the advent of scientific agriculture, involve improvement in absolute terms that can be applied to all manners of producing corn. Not all the innovations even involve a specific improvement, appropriate for concrete, historical conditions.

None of the techniques that we are familiar with are pure. There is a constant flow of elements between productive systems that is much more common than the transfer of complete systems. Innovations have many sources. It would be the height of arrogance to credit scientific agriculture with the role of creator of and primary contributor to such innovations. Scientific agriculture has the best documented history. It is precisely such data that allows us to establish its interdependence with respect to other productive systems as sources of innovation. I only wish that we possessed a similar body of information on other agricultural traditions. Still more arrogant is the identification of scientific agriculture as a unique system of production, the one true way. That false identity, a burdensome legacy of the Green Revolution, emphasizes one of the most severe and obvious limitations on agricultural research: its concentration on a narrow repertoire of objectives and conditions of production. Even so, the application of modern scientific investigation has been an important element in the increase of diversity through the adoption of some of its elements on the part of many systems of production. I hope that in the future it also is due to its application to all ways of farming, an objective that still remains elusive.

Corn is becoming more and more diverse in the ways in which it is consumed. Its use as fodder and forage in cattle fattening and the growth of the refining industry since the advent of wet grinding stand out as substantive changes by virtue of their scale and magnitude. Another significant development has been the growing market for processed corn, almost always but not exclusively destined for foodstuffs. Corn as the raw material for these products has lent this plant an elasticity of demand not typical of the cereal market in general. Changes such as these have not supplanted traditional uses for corn in which this grain enters into diets directly as a staple food. Never before have more people depended on corn for their immediate and direct physical survival.

Corn's traditional uses have not remained unchanged. There probably is no better example of this than the complex industry that has arisen in modern Mexico to provide the population with freshly made tortillas using ready-made tortilla dough. In 1980, nearly 61,500 persons worked either in mills producing ready-made dough or in tortilla factories (Museo de Culturas Populares, 1982a: 83), four times the 15,000 workers directly employed by the twenty-three corn-refining plants in the United States. Despite the magnitude and complexity of this industrial network, the traditional use of the corn tortilla as the mainstay of the Mexican diet has remained unchanged. It is not possible to discern any linearity in the diverse patterns of corn consumption. All change in their own way, without losing their particular style, their appearance. No single norm has been established as desirable for those consumers who eat corn and use it to their liking. The model of U.S. consumption is not only extravagant in its use of energy, making it inappropriate on a world scale, but is not attractive to those who consume corn in other ways. Nothing is more false than the supposition that hunger and malnutrition are eradicated by substituting the consumption of wheat bread for corn, an assumption inevitably reminiscent of Marie Antoinette and her infamous cake.

The greatest source of diversity lies in the political, social, and material conditions that separate groups and classes of corn producers and corn consumption worldwide. It is not uncommon for individual average corn consumption to vary in a ration of five to one. These differences are exacerbated because those who consume less are more strictly dependent on corn as a staple food, receiving up to two-thirds of all their nutrients from that grain. They also are made worse because those who consume less but depend more on corn for their nutritional intake do not end up producing corn in sufficient quantity to provide for the population at large. Per capita income disparity ranging from one to one hundred is not uncommon. Neither are the thirty or more years of difference in average life expectancies remarkable. Those statistical expressions that at times emphasize differences also often mask them.

Brief Reflections

Many times, the higher the figure, the higher the waste and irrationality. The difference is enormous, even though there is no precise yardstick by which to measure it. Worse still, the difference is an abyss that grows bigger from one day to the next. There is a cruel irony in so many empty promises. Despite all the predictions and theories that wealth will spill over from above, first as a trickle and later as an unrestrained waterfall, the base of the pyramid is becoming ever wider and its vertex ever more narrow.

Inequality is the form of diversity that is most directly and narrowly linked to the spread of capitalism. Exploitation, unequal exchange, transfer of surpluses, are all analytical concepts particular to the essence of capitalism. The cause and object of capitalist expansion is the private accumulation of capital, of wealth that possesses the ability to reproduce itself. This is not a characteristic inherent to wealth, nor even to accumulation. Every social system has forms of accumulation and differentials of wealth distinguishing its constituent members. Only in some social systems is social accumulation privatized and only in capitalism can wealth transform itself into capital upon acquiring the capacity to purchase the means of production and labor. Capital is not derived from nature, but from society. Capitalist society establishes the relations that make the private appropriation of the means of production and the purchase of labor as a commodity feasible. Those are relations of politics and of power.

Strictly speaking, the transformation of wealth into capital only takes place in the capitalist mode of production. A historical analysis opens up other perspectives. Here, the distinction between mercantile wealth and capital become difficult to make and the overlapping of these with imperialism is continual. Capitalism does not expand thanks to any embedded inherent economic advantages that pertain. Rather, capitalism expands fundamentally through the use of military force and political domination. Political domination is used to impose the relations that permit the transformation of resources and products of labor into capital through uneven exchange with the metropolis. The formation of capital historically unfolds on a world stage, but its accumulation is concentrated in a few centers and classes. Political borders, national boundaries, separate and protect the ubiquitous production of surpluses and their transformation into capital.

Under those circumstances, capitalist exploitation did not result in reproduction of the internal norms of the capitalist mode of production, but in the appearance of new forms that did not repeat the past. African slavery on the American continents never recreated the slavery of classical antiquity. The African slave trade did not faithfully reproduce any previous experience with slavery, but was derived directly from the expansion and domination of capitalism. That took place at the same time and together with the proletarianization of workers and the Industrial Revolution, with the destruction of latifundio and with its creation, with the development of new servile relations and with the appearance of owner-operators. Sharecropping, furnishing, ground rent paid to absentee landowners, these were not simply the tools of capitalist interests. They were transformed until they bore little resemblance to their ancestral forms. The persistence of what were perceived to be precapitalist forms was often a simple nominal illusion. In appearance the inertia of the past continued to prevail. In reality, those forms underwent radical transformations, such as those suffered by salaried workers. The formation of capitalism was from the start a worldwide process, diversified, multiple, heterogeneous. That was how capitalism developed; that is the form of capitalism today.

For the majority of humanity, contact with and submission to capitalism had a political character. The relation of force so necessary to capitalism had an extraeconomic nature. Political coercion never disappeared. Political coercion mingled with economic motives and interests. These never acquired the purity attributed to them by abstract theoretical models. The strictly economic attraction of capitalism was greatly exaggerated. The comparative advantages, the technological superiority, the greater efficiency attributed to capitalist production only came about under a system of coercion and political domination. Historically, capitalism could not be analyzed in isolation or primarily as an economic phenomenon. Such a perspective arose in and emanated from the centers of accumulation, where domination materialized as capital and as wealth. Elsewhere, domination engendered poverty, the constant expansion and increasingly demanding extraeconomic coercion of poverty through authoritarian rule. These different perspectives were valid, with none being absolute or constituting a universal model. Dominance of one perspective over another was determined by jockeying for position, by unfolding processes, by concrete forces. Economics and politics could and should be separated for the purposes of historical analysis, but they need to be reunited in the search for explanations.

Intrinsic inequality is a diversifying force of enormous importance and proportion. Inherent inequity is a structural factor that is part of the capitalist mode of production and its associated political domination. That it exists at all belies the promise of uniformity in the capitalist millennium. If intrinsic inequality were our only means to explain diversity, then other forces that are a part of its structure would be overlooked. Other social, technical, and material relations contribute to the structure of diversity in many ways. Different trends and norms in civilizations and cultures often have opposed or resisted capitalist domination and have obliged the dominators and the dominated to undergo transformation instead. They also have submitted to a prevailing trend, but on their own terms and in novel forms. The history of capitalism is a history of triumph, but not of omnipotence. The contamination of a theoretical supposition is much more frequent than its preservation in a pure form. It can be postulated that capitalism has been transformed more by external pressures and resources than by its endogenous dynamic. Corn's complex history worldwide suggests this. The same postulate is valid for those who were dominated. The interaction is what would generate a greater transformational force.

Capitalism was constructed by interaction on a world scale and not by manifest destiny. Its construction was a ubiquitous phenomenon. It happened simultaneously the world over in different ways. While capitalism's route may not have been unambiguous and uncluttered, it did have the reproduction of capital to guide it. The slave, the sharecropper, and the free agriculturist contributed their share to that process right along with the new industrial laborer or the scientific worker, the inventor. That process saw all of them and many more at once classified, sorted into specific classes, and thrown together. The face-off between two fundamental classes never took place historically, although the idea persisted ideologically. Classes multiplied without final convergence. The forms of exploitation became more diverse and generated multiple contradictions. Complex alliances and heterogeneous coalitions played decisive roles in political transformations with no foregone conclusions.

It is surprising how what historically was a process of articulated diversity today is seen instead as a simple sequence, as an inevitable sort of domino effect, largely intolerant of any gaps of the type suggested by radical postulates. The same sort of reasoning that conceptually converted an integrated reality that was diverse, varied, and interactive into a rigid sequential process dangles capitalism before us as an incentive and promise of the future. The advent of capitalism already is a historical fact, a part of the past. Capitalist integration already has taken place. It already is part of our shared experience, our collective memory, and, occasionally, our accrued rancor. We are promised the past. We are told that we will all be princes, overlords, or modest and honorable proletarians far removed from hunger and uncertainty, from ignominy. It has not happened thus far and it is not clear why it should happen now.

The same slight of hand that promotes the past as the future has inculcated the idea in us that capitalism is something foreign to the Third World, a patrimony of the First World alone, of the wealthy, of others, of the knowledgeable, of the capable. It converts a specific difference of position into a historical distance. The poor are not recognized as such, but are explained as those who are left behind. That which is different is explained as backward and marginal. Those who are poor and different are deemed responsible for creating their own circumstances. Their contribution to the construction of capitalism is degraded in order to accentuate the importance of those who hold wealth. That which separates them is a relation of power, not of superiority. The concept of backwardness inhibits and oppresses them. It distances them from their own history, from a story that already has been repeated several times over. The condemnation of precapitalist forms dismisses their experience in order to offer them something they already are familiar with, something they already have lived through. It sounds like heresy, but capitalism is as much their legacy as it is the legacy of international bankers. Capitalism, understood in its broadest sense, has no historical owner. Capital, power, and, in many senses, history do.

It is in this sense, the ownership of history, that this work attempts to make an impact. It is not everything, but it certainly is something. History is, among other things, the only basis from which to speculate about the future, to discern laws and probabilities, to construct desirable or catastrophic scenarios. Among the conceptions about the future, I distinguish between millennia, futures imposed from without by inevitable laws or divine design, and utopias, expressions of desire, of demands and struggles, projections of experience and of forward-looking history. Neither millennia nor utopias are probable, just as no prediction about the future can be. But both can be real and realistic as guides to our routines, to our daily struggles. The capitalist millennium is already upon us, its promises already kept. They are part of our memory, part of what we already know. The distance between its promises and its fulfillment is abysmal. It is better not to look at this chasm as defeat or as time running out. Rather, this breach only implies that the promises are false, that they are the products of laws that are simply untrue.

Perhaps now is the time to propose a rejection of a new millennium, of a prepackaged future, in order to rethink utopias as an open invitation for debate over desires and demands, free from determinism or destiny. It seems to me that the past was constructed in just that way. What happened simply was; thus forces and circumstances came together in a certain way, but that course was not the only possible outcome. To insist on inevitability is merely to foretell that which has happened without understanding it. Other things might happen. The future can be this way, wide open, subject to the struggle and agreement between utopias and millennia in order to structure forces and construct circumstances.

The above inevitably sounds volunteristic. It is not entirely so. If history rids itself of inexorable and inflexible laws, if history refuses to recognize purely determinant factors, then yes, areas, forces, and conditions appear that form a chain reaction that acquires more weight, more capacity for transformation. These critical factors, as those that stand out in this work, perhaps are not the only ones, but they are a beginning. We do not know everything about them, but neither are we totally untutored. This book deals with limited processes: relations between plants and people, the trauma and intrigue between natural processes and social processes, their linkage and autonomy in conditions of subordination of those who produce food for other sectors of society. I hope that I have successfully established what some of those critical factors are. This is not enough. There is still much to do.

One of the many problems for such a necessary and possible task is the lack of historical accounts that might enrich our knowledge about the linkages critical to such transformation. Almost all conventional histories were written from a position of triumph and power. We know much more about domination than about resistance. We are better acquainted with a few select minorities than with all the rest. For this reason, we frequently go wrong and we derive universal laws that are not universal after all. Too many gaps and silences speak volumes. This book contributes to filling in some of those gaps. It attempts to attribute importance to processes and topics that have not had the benefit of analysis in the past. This book attempts to take history off its pedestal and move toward the collective and the customary, a little closer to shared experience, to the quotidian. Its highest ambition would be that of humbly working in the service of utopias as I understand them, building a future based on what is near at hand, on diversity, on intimacy; on that which should be preserved, improved, and become more familiar. Building such a future depends on transforming whatever impedes or threatens such an enterprise. For that reason the protagonist of the future is a common but marvelous plant, scarcely fit to eat, a bastard to boot, the prodigious inheritance of nameless multitudes.

BIBLIOGRAPHY

Abel, Wilhelm. 1986. *La agricultura: Sus crisis y coyunturas: Una historia de la agricultura y la economía alimentaria en Europa Central desde la Alta Edad Media*. Mexico: Fondo de Cultural Económica.

Acosta, José de. 1940. *Historia natural y moral de las Indias*. Mexico: Fondo de Cultura Económica.

Aguirre Beltrán, Gonzalo. 1972. *La población negra de México*. Mexico: Fondo de Cultura Económica.

Aiton, Arthur S. 1951. "The Impact of the Flora and Fauna of the New World upon the Old World during the Sixteenth Century." *Chronica Botanica* 12, nos. 4–6:121–25.

Allan, W. 1971. "The Normal Surplus of Subsistence Agriculture." In *Economic Development and Social Change: The Modernization of Village Communities*, edited by George Dalton, 88–98. Garden City, N.Y.: Natural History Press.

Allen, R. L. 1853. *The American Farm Book: or Compend of American Agriculture*. New York: C. M. Saxton.

American Plant Studies Delegation, Committee on Scholarly Communication with the People's Republic of China. 1975. *Plant Studies in the People's Republic of China: A Trip Report of the American Plant Studies Delegation, Submitted to the Committee on Scholarly Communication with the People's Republic of China*. Washington, D.C.: National Academy of Sciences.

Amicus Curiae. 1847. *Food for the Million: Maize against Potato, a Case for the Times, Comprising the History, Uses and Culture of Indian Corn, and Especially Showing the Practicability and Necessity of Cultivating, the Dwarf Varieties, in England and Ireland*. London: Longman, Brown, Green, and Longman.

Amin, Samir. 1976. *Unequal Development*. New York: Monthly Review Press.

Anderson, Edgar. 1952. *Plants, Man, and Life*. Boston: Little, Brown.

Anderson, Jeremy. 1967. "A Historical-Geographical Perspective on Khrushchev's Corn Program." In *Soviet and East European Agriculture*, edited by Jerzy F. Karcz, 104–28. Berkeley: University of California Press.

Anes Álvarez, Gonzalo, et al. 1978. *La economía agraria en la historia de España: Propiedad, explotación, rentas*. Madrid: Ediciones Alfaguara.

Anglería, Pedro Mártir de. 1964–65. *Las décadas del Nuevo Mundo*. Mexico: Porrúa.

Ardrey, Robert L. 1900. *American Agricultural Implements: A Review of Invention and Development in the Agricultural Implement Industry of the United States*. Chicago: n.p.

Aykroyd, W. R., I. Alexa, and J. Nitzulescu. 1981. "Study of the Alimentation of Peasants in the Pellagra Area of Moldavia (Roumania)." In *Pellagra*, edited by Kenneth J. Carpenter, 31–46. Stroudsburg, Pa.: Hutchinson Ross.

Badell y Abadía, Gabriel García. 1963. *Introducción a la historia de la agricultura española*. Madrid: Consejo Superior de Investigaciónes Científicas.

Bailey, Liberty H., ed. 1907, 1909. *Cyclopedia of American Agriculture: A Popular Survey of Agricultural Conditions, Practices, and Ideals in the United States and Canada*. New York: Macmillan.

Barkin, David, and Blanca Suárez. 1981. *El complejo de granos en México*. Mexico: Centro de Ecodesarrollo e Instituto Latinamericano de Estudios Transnacionales.

Barlow, Joel. 1853. "The Hasty Pudding: A Poem in Three Cantos. Written at Chamrery in Savoy, January 1793 . . . With a Memoir on Maize or Indian Corn Compiled by D. J. Browne, under the Direction of the American Institute." In *The American Farm Book*, edited by R. L. Allen. New York: C. M. Saxton.

Bates, Robert H. 1981. *Markets and States in Tropical Africa: The Political Basis of Agricultural Policies*. Berkeley: University of California Press.

Beadle, George W. 1978. "Teosinte and the Origin of Maize." In *Maize Breeding and Genetics*, edited by David B. Walden, 113–28. New York: Wiley.

———. 1982. "El origen del maíz comprobado por medio del polen." *Información Científica y Tecnológica* 4, no. 72.

Bement, C. N. 1854. "History of Indian Corn: Its Origin, Its Culture, and Its Uses." In *Transactions of the N.Y. State Agricultural Society* 13 (1853): 321–53. Albany: C. Van Benthuysen.

Benítez, Fernando. 1986. *Ki: El drama de un pueblo y de una planta*. Mexico: Fondo de Cultura Económica.

Bennett, Merrill K., and Rosamond H. Pierce. 1961. "Change in the American National Diet, 1879–1959." *Food Research Institute Studies* 2, no. 2:95–119.

Bignardi, Agostino. 1983. *Disegno storico dell' agricoltura italiana*. Bologna: Li Causi.

Birket-Smith, Kaj. 1943. *The Origin of Maize Cultivation*. Copenhagen: I Kommission Los E. Munksgaard.

Bitterli, Urs. 1989. *Cultures in Conflict: Encounters between European and Non-European Cultures, 1492–1800*. Translated by Ritchie Robertson. Cambridge: Polity.

Blunt, Wilfrid, and Sandra Raphael. 1979. *The Illustrated Herbal*. New York: Thames and Hudson.

Bogue, Allan G. 1963. *From Prairie to Corn Belt: Farming on the Illinois and Iowa Prairies in the Nineteenth Century*. Chicago: University of Chicago Press.

Bohannan, Paul, and Philip Curtin. 1971. *Africa and Africans*. Garden City, N.Y.: Natural History Press.

Bonafous, Matthieu. 1836. *Histoire naturelle, agricole et économique du maïs*. Paris: Huzard.

Boserup, Ester. 1965. *The Conditions of Agricultural Growth: The Economics of Agrarian Change under Population Pressure*. New York: Aldine.

Bowden, Witt, Michael Karpovich, and Abbott Payson Usher. 1970. *An Economic History of Europe since 1750*. New York: AMS Press.

Bowman, Melville LeRoy. 1915. *Corn: Growing, Judging, Breeding, Feeding, Marketing*. Waterloo, Iowa: n.p.

Braudel, Fernand P. 1975. *The Mediterranean and the Mediterranean World in the Age of Philip II*. New York: Harper and Row.

——. 1981. *Civilization and Capitalism, 15th–18th Centuries*. New York: Harper and Row.

Braudel, Fernand P., and F. Spooner. 1967. "Prices in Europe from 1450–1750." In *The Cambridge Economic History of Europe*, vol. 4, *The Economy of Expanding Europe in the Sixteenth and Seventeenth Centuries*, edited by Edwin E. Rich and C. H. Wilson, 374–486. Cambridge: Cambridge University Press.

Brooks, Eugene C. 1916. *The Story of Corn and the Westward Migration*. Chicago: Rand McNally.

Brown, Lester R. 1971. *Man, Land, and Food: Looking Ahead at World Food Needs*. Washington, D.C.: U.S. Department of Agriculture, Economic Research Service, Regional Analysis Division.

Brown, Lester R., with Erik P. Eckholm. 1974. *By Bread Alone*. New York: Praeger.

Bruman, Henry. 1937. "The Russian Investigations on Plant Genetics in Latin America and Their Bearing on Culture History." In *Handbook of Latin American Studies for 1936*, edited by Lewis Hanke, 449–58. Cambridge: Harvard University Press.

Buck, John L. 1930. *Chinese Farm Economy*. Chicago: University of Chicago Press.

——. 1937. *Land Utilization in China*. Nanking: University of Nanking.

Buniva, Prof. 1981. "Observations on Pellagra: It Would Not Appear to Be Contagious." In *Pellagra*, edited by Kenneth J. Carpenter, 11–12. Stroudsburg, Pa.: Hutchinson Ross.

Bunting, E. S. 1968. "Maize in Europe." *Field Crop Abstracts* 21, no. 1:1–9.

Burbach, Roger, and Patricia Flynn. 1980. *Agribusiness in the Americas*. New York: Monthly Review Press.

Burtt-Davy, Joseph. 1914. *Maize: Its History, Cultivation, Handling, and Uses, with Special Reference to South Africa*. London: Longman, Green.

Butz, Earl L. 1982. "El maíz, clave de la opulencia." *El Gallo Ilustrado*, weekly supplement to *El Día*, no. 1020 (January 3).

Candolle, Alphonse de. 1902. *Origin of Cultivated Plants*. New York: Appleton.

Cárdenas, Juan de. 1988. *Problemas y secretos maravillosos de las Indias*. Madrid: Alianza Editorial.

Carpenter, Kenneth J., ed. 1981. *Pellagra*. Stroudsburg, Pa.: Hutchinson Ross.

Carrera Pujal, Jaime. 1947. *Historia de la economía española*. Barcelona: Bosch.

Carrier, Lyman. 1923. *The Beginnings of Agriculture in America*. New York: McGraw-Hill.

Carter, George F. 1945. *Plant Geography and Culture History in the American Southwest*. New York: Viking Fund Publications in Anthropology.

Casas, Bartolomé de las. 1951. *Historia de las Indias*. Mexico: Fondo de Cultura Económica.

Chang, K. C., ed. 1977. *Food in Chinese Culture: Anthropological and Historical Perspectives*. New Haven: Yale University Press.

Christianson, D. D., J. S. Wall, R. J. Dimler, and A. N. Booth. 1981. "Nutritionally Unavailable Niacin in Corn: Isolation and Biological Activity." In *Pellagra*, edited by Kenneth J. Carpenter, 311–15. Stroudsburg, Pa.: Hutchinson Ross.

Cipolla, Carlo M. 1962. *The Economic History of World Population*. Baltimore: Penguin Books.

——. 1976. *Before the Industrial Revolution: European Society and Economy, 1000–1700*. New York: Norton.

Clark, William H. 1945. *Farms and Farmers: The Story of American Agriculture*. Boston: L. C. Page.

Clout, Hugh D. 1977. "Agricultural Change in the Eighteenth and Nineteenth Centuries." In *Themes in the Historical Geography of France*, edited by Hugh D. Clout, 407–46. New York: Academic Press.

Cobbett, William. 1828. *A Treatise on Cobbett's Corn*. London: W. Cobbett.

Cochrane, Willard W. 1979. *The Development of American Agriculture: A Historical Analysis*. Minneapolis: University of Minnesota Press.

Códice Chimalpopoca. 1975. Translated by Primo Feliciano Velázquez. Mexico: Universidad Nacional Autónoma de México.

Cole, John N. 1979. *Amaranth, from the Past to the Future*. Emmaus, Pa.: Rodale Press.

Collins, Guy N. 1919. "Notes on the Agricultural History of Maize." In *Annual Report of the American Historical Association for the Year 1919* 1, no. 8:409–29. Washington, D.C.: Government Printing Office.

Consejo Nacional de Ciencia y Tecnología, Centro de Investigación en Química Aplicada, Comisión Nacional de las Zonas Çridas. 1978. *Guayule, reencuentro con el desierto*. Mexico: Consejo Nacional de Ciencia y Tecnología.

Cook, O. F. 1932. "The Debt of Agriculture to Tropical America." In *Annual Report of the Board of Regents of the Smithsonian Institution Showing the Operations, Expenditures, and Condition of the Institution for the Year Ending June 30, 1931*, 491–501. Washington, D.C.: Government Printing Office.

Cooper, Frederick. 1981. "Africa and the World Economy." *African Studies Review* 24, nos. 2–3:1–86.

Corn Refiners Association, Inc. n.d. *Amazing Maize: Background Memorandum*. Washington, D.C.: n.p.

Cox, George W., and Michael D. Atkins. 1979. *Agricultural Ecology: An Analysis of World Food Production Systems*. San Francisco: W. H. Freeman.

Cravioto, René O., et al. 1945. "Nutritive Value of the Mexican Tortilla." *Science* 102, no. 2639 (July 27): 91–93.

Crosby, Alfred W., Jr. 1972. *The Columbian Exchange: Biological and Cultural Consequences of 1492*. Westport, Conn.: Greenwood Press.

Cummings, Richard Osborn. 1940. *The American and His Food: A History of Food Habits in the United States*. Chicago: University of Chicago Press.

Curtin, Philip D. 1969. *The Atlantic Slave Trade: A Census*. Madison: University of Wisconsin Press.

———. 1984. *Cross-Cultural Trade in World History*. Cambridge: Cambridge University Press.

Cutler, Hugh C. 1968. "Origins of Agriculture in the Americas." *Latin American Research Review* 3, no. 4 (Fall): 3–21.

Cutler, Hugh C., and Leonard W. Blake. 1971. "Travels of Corn and Squash." In *Man across the Sea*, edited by Carroll L. Riley et al., 366–75. Austin: University of Texas Press.

Cutler, Hugh C., and Martín Cárdenas. 1947. "Chicha, a Native South American Beer." *Botanical Museum Leaflets, Harvard University* 13, no. 3 (December 29): 33–60.

Dalton, George, ed. 1971. *Economic Development and Social Change: The Modernization of Village Communities*. Garden City, N.Y.: Natural History Press.

Davidson, Basil. 1980. *The African Slave Trade*. Boston: Little, Brown.

Davis, Ralph. 1973. *The Rise of the Atlantic Economies*. Ithaca, N.Y.: Cornell University Press.

Deane, Phyllis. 1965. *The First Industrial Revolution*. Cambridge: Cambridge University Press.

Dias, Jorge, Ernesto Veiga de Oliveira, and Fernando Galhano. 1961. *Sistemas primitivos de secagem e armazenagem de produtos agrícolas: Os espigueiros portugueses*. Porto, Portugal: Instituto de Alta Cultura.

Dinham, Barbara, and Colin Hines. 1984. *Agribusiness in Africa*. London: Africa World Press.

Dovring, Folke. 1965. "The Transformation of European Agriculture." In *The Cambridge Economic History of Europe*, vol. 6, pt. 2, *The Industrial Revolutions and After: Income, Population, and Technological Change*, edited by H. J. Habakkuk and M. Postan, 603–72. Cambridge: Cambridge University Press.

Dressler, Robert L. 1953. "The Pre-Columbian Cultivated Plants of Mexico." *Botanical Museum Leaflets, Harvard University* 16, no. 6:115–72.

Duncan, W. G. 1975. "Maize." In *Crop Physiology: Some Case Histories*, edited by L. T. Evans, 23–50. London: Cambridge University Press.

Ebeling, Walter. 1979. *The Fruited Plain: The Story of American Agriculture*. Berkeley: University of California Press.

Elliot, J. H. 1970. *The Old World and the New, 1492–1650*. Cambridge: Cambridge University Press.

Enfield, Edward. 1866. *Indian Corn: Its Value, Culture, and Uses*. New York: D. Appleton.

Erwin, A. T. 1950. "The Origin and History of Pop Corn." *Economic Botany* 4, no. 3:294–99.

Etheridge, Elizabeth. 1972. *The Butterfly Caste: A Social History of Pellagra in the South*. Westport, Conn.: Greenwood Press.

Fage, J. D. 1969. *A History of West Africa: An Introductory Survey*. London: Cambridge University Press.

Fallers, L. A. 1971. "Are African Cultivtors to Be Called 'Peasants.'" In *Economic Development and Social Change: The Modernization of Village Communities*, edited by George Dalton, 169–77. Garden City, N.Y.: Natural History Press.

Fausz, J. Frederick. 1981. "Opechancanough: Indian Resistance Leader." In *Struggle and Survival in Colonial America*, edited by David G. Sweet and Gary B. Nash, 21–37. Berkeley: University of California Press.

Fél, Edit, and Tamás Hofer. 1969. *Proper Peasants: Traditional Life in a Hungarian Village*. New York: Wenner-Gren Foundation for Anthropological Research.

Felloni, Giuseppe. 1977. "Italy." In *An Introduction to the Sources of European Economic History, 1550–1800*, edited by Charles Wilson and Geoffrey Parker, 1–36. Ithaca, N.Y.: Cornell University Press.

Fernández de Oviedo, Gonzalo. 1978. *Historia general y natural de las Indias*. Mexico: Centro de Estudios de Historia de México.

Finan, John J. 1950. *Maize in the Great Herbals*. Waltham, Mass.: Chronica Botanica.

Fite, Gilbert. 1979. "Southern Agriculture since the Civil War: An Overview." *Agricultural History* 53, no. 1:3–21.

Fogel, Robert William, and Stanley L. Engerman. 1974. *Time on the Cross: The Economics of American Negro Slavery*. Boston: Little, Brown.

Food and Agriculture Organization of the United Nations. 1968. *FAO Production Yearbook*. Rome: Food and Agriculture Organization of the United Nations.

———. 1973. *FAO Production Yearbook*. Rome: Food and Agriculture Organization of the United Nations.

———. 1983. *FAO Production Yearbook*. Rome: Food and Agriculture Organization of the United Nations.

Fornari, Harry. 1973. *Bread upon the Waters: A History of United States Grain Exports*. Nashville: Aurora Publishers.

Fried, Morton H. 1953. *Fabric of Chinese Society: A Study of the Social Life of a Chinese Country Seat*. New York: Praeger.

Frundianescu, A., and G. Ionescu-Sisesti. 1933. "Aspects of Rumanian Agriculture." In *Agricultural Systems of Middle Europe*, edited by Ora S. Morgan, 307–52. New York: Macmillan.

Funk, Casimir. 1981. "Studies on Pellagra: The Influence of the Milling of Maize on the Chemical Composition and the Nutritive Value of Maize-Meal." In *Pellagra*, edited by Kenneth J. Carpenter, 316–19. Stroudsburg, Pa.: Hutchinson Ross.

Fussell, G. E. 1966. "Ploughs and Ploughing before 1800." *Agricultural History* 40, no. 3 (July): 177–86.

———. 1972. *The Classical Tradition in West European Farming*. Rutherford, N.J.: Fairleigh Dickinson University Press.

Galinat, Walton C. 1965. "The Evolution of Corn and Culture in North America." *Economic Botany* 19, no. 4:350–57.

García Lombardero, Jaime. 1974. "Aportación al estudio del sector agrario en la Galicia del siglo XVIII: Un contraste con Cataluña." In *Agricultura, comercio colonial y crecimiento económico en la España contemporánea*, edited by Jordi Nadal and Gabriel Tortella, 44–66. Barcelona: Editorial Ariel.

Gates, Paul W. 1965. *Agriculture and the Civil War*. New York: Knopf.

Genevois, L. 1961. "Le maïs et la pellagre." *Journal d'Agriculture Tropicale et de Botanique Appliquée* 8, nos. 6–7 (June, July): 221–28.

Genovese, Eugene D. 1967. *The Political Economy of Slavery: Studies in the Economy and Society of the Slave South*. New York: Pantheon.

———. 1976. *Roll, Jordan, Roll: The World That Slaves Made*. New York: Vintage Books.

George, Susan. 1977. *How the Other Half Dies: The Real Reasons for World Hunger*. Montclair, N.J.: Allanheld, Osmun.

Gerbi, Antonello. 1973. *The Dispute of the New World: The History of a Polemic, 1750–1900*. Pittsburgh: University of Pittsburgh Press.

———. 1985. *Nature in the New World: From Christopher Columbus to Gonzalo Fernandez de Oviedo*. Translated by Jeremy Moyle. Pittsburgh: University of Pittsburgh Press.

Godinho, Vitorino Magalhães. 1984. *Os descobrimentos e a economia mundial*. Lisboa: Editoral Presença.

Gómez Tabanera, José M. 1973. "En torno a la introducción en Europea del *Zea mays* y su adopción por asturias y el noroeste hispánico." *Boletín del Instituto de Estudios Asturianos* 27, no. 78:157–201.

Gourou, Pierre. 1959. *Los países tropicales*. Xalapa: Universidad Veracruzana.

Grigg, David B. 1974. *The Agricultural Systems of the World: An Evolutionary Approach.* London: Cambridge University Press.

———. 1980. *Population Growth and Agrarian Change: An Historical Perspective.* Cambridge: Cambridge University Press.

Guyer, Jane I. 1981. "Household and Community in African Studies." *African Studies Review* 24, nos. 2–3:87–137.

Harasti da Buda, Gaetano. 1788. *Della coltivazione del maiz: Memoria che riportò il premio [dell' accessit] dalla Pubblica Accademia Agraria di Vicenza nel di 2 Ottobre 1786.* Vicenza: Nella Stamperia Turra.

Hardeman, Nicholas P. 1981. *Shucks, Shocks, and Hominy Blocks: Corn as a Way of Life in Pioneer America.* Baton Rouge: Louisiana State University Press.

Harlan, Jack R. 1975. *Crops and Man.* Madison: American Society of Agronomy.

———. 1976. "The Plants and Animals That Nourish Man." *Scientific American* 235, no. 3:88–97.

Harris, Henry F. 1919. *Pellagra.* New York: Macmillan.

Harrison, John A. 1967. *China since 1800.* New York: Harcourt, Brace and World.

Harshberger, J. W. 1893. "Maize: A Botanical and Economic Study." *Contributions from the Botanical Laboratory of the University of Pennsylvania* 1, no. 2:75–303.

Hedrick, Ulysses Prentiss. 1966. *A History of Agriculture in the State of New York.* New York: Hill and Wang.

Heiser, Charles B., Jr. 1973. *Seed to Civilization: The Story of Man's Food.* San Francisco: W. H. Freeman.

Helleiner, Karl F. 1967. "The Population of Europe from the Black Death to the Eve of the Vital Revolution." In *The Cambridge Economic History of Europe*, vol. 4, *The Economy of Expanding Europe in the Sixteenth and Seventeeenth Centuries*, edited by Edwin E. Rich and C. H. Wilson, 1–95. Cambridge: Cambridge University Press.

Hemardinquer, Jean-Jacques. 1963. "L'Introduction du maïs et la culture des sorghos." *Bulletin Philologique et Historique* 1:429–59.

Hernández, Francisco, and Germán Somolinos d'Ardois. 1960–84. *Obras completas.* Mexico: Universidad Nacional Autónoma de México.

Hernández Xolocotzi, Efraim. 1985. *Xolocotzia: Obras de Efraim Hernández Xolocotzi.* Mexico: Universidad Autónoma Chapingo, Revista de Geografía Agrícola.

Ho, Ping-ti. 1955. "The Introduction of American Food Plants in China." *American Anthropologist* 57, no. 2:191–201.

———. 1959. *Studies on the Population of China, 1368–1953.* Cambridge: Harvard University Press.

Hobhouse, Henry. 1986. *Seeds of Change: Five Plants That Transformed Mankind.* New York: Harper and Row.

Hommel, Rudolf P. 1969. *China at Work.* Cambridge: MIT Press.

Hopkins, A. G. 1973. *An Economic History of West Africa.* London: Longman Group.

Humboldt, Alexander von. 1972. *Political Essay on the Kingdom of New Spain*. Translated by John Black. Edited by Mary Maples Dunn. New York: Knopf.

Iltis, Hugh H. 1983. "From Teosinte to Maize: The Catastrophic Sexual Transmutation." *Science* 222, no. 4626 (November 25): 886–94.

Jeffreys, M. D. W. 1971. "Pre-Columbian Maize in Asia." In *Man across the Sea*, edited by Carroll L. Riley et al., 376–400. Austin: University of Texas Press.

Johnson, Charles S., Edwin R. Embree, and W. W. Alexander. 1935. *The Collapse of Cotton Tenancy: Summary of Field Studies and Statistical Surveys, 1933–35*. Chapel Hill: University of North Carolina Press.

Johnston, Bruce F. 1958. *The Staple Food Economies of Western Tropical Africa*. Stanford: Stanford University Press.

Johnston, Bruce F., and Peter Kilby. 1975. *Agriculture and Structural Transformation: Economic Strategies in Late-Developing Countries*. New York: Oxford University Press.

Jones, Robert H. 1963. "Long Live the King?" *Agricultural History* 37, no. 3:166–69.

July, Robert W. 1974. *A History of the African People*. New York: Scribner.

Kahn, E. J., Jr. 1984. "Profiles: The Staffs of Life—The Golden Thread." *New Yorker*, June 18, 46–88.

Kalm, Peter. 1974. "Description of Maize: How It Is Planted and Cultivated in North America and the Various Uses of This Grain." *Economic Botany* 28, no. 2:105–17.

Kemmerer, Donald L. 1949. "The Pre–Civil War South's Leading Crop: Corn." *Agricultural History* 23, no. 4:236–39.

Kempton, James Howard. 1931. "Maize: The Plant-Breeding Achievement of the American Indian." In *Old and New Plant Lore: A Symposium*, 11:319–49. New York: Smithsonian Institution Series.

———. 1938. "Maize: Our Heritage from the Indian." In *Annual Report of the Board of Regents of the Smithsonian Institution Showing the Operations, Expenditures, and Condition of the Institution for the Year Ending June 30, 1937*, 385–408. Washington, D.C.: Government Printing Office.

Kodicek, E., R. Braude, S. K. Kon, and K. G. Mitchell. 1981. "The Effect of Alkaline Hydrolysis of Maize on the Availability of Its Nicotinic Acid to the Pig." In *Pellagra*, edited by Kenneth J. Carpenter, 293–310. Stroudsburg, Pa.: Hutchinson Ross.

Kuleshov, N. N., Harold J. Kidd, and Howard C. Reynolds. 1954. "Some Peculiarities of Maize in Asia." *Annals of the Missouri Botanical Garden* 41, no. 3 (September): 271–99.

Kupzow, A. J. 1967–68. "Histoire du maïs." *Journal d'Agriculture Tropicale et de Botanique Appliquée* 14, no. 12:526–61, and 15, nos. 1–3:42–68.

Langer, William L. 1963. "Europe's Initial Population Explosion." *American Historical Review* 69, no. 1:1–17.

Lappé, Francis Moore, and Joseph Collins. 1977. *Food First: Beyond the Myth of Scarcity*. Boston: Houghton-Mifflin.

Laufer, Berthold. 1906. "The Introduction of Maize into Eastern Asia." *International Congress of Americanists* 15 (September 15): 223–57.

Lavinder, Charles H. 1912a. "The Association for the Study of Pellagra: A Report of the Second Triennial Meeting Held at Columbia, S.C., October 3–4, 1912." *Public Health Reports* 27, pt. 2, no. 44 (November 1): 1776–78.

——. 1912b. "Pellagra in Italy: A Note on the Prevalence in 1881, 1899, 1910." *Public Health Reports* 27, pt. 2, no. 44 (November 1): 1778–79.

——. 1912c. "The Prevalence and Geographic Distribution of Pellagra in the United States." *Public Health Reports* 27, pt. 2, no. 50 (December 13): 2076–88.

Leach, Edward R. 1954. *Political Systems of Highland Burma: A Study of Kachim Social Structure*. Cambridge: Harvard University Press.

Long-Solís, Janet. 1986. *Capsicum y cultura: La historia del chili*. Mexico: Fondo de Cultura Económica.

Lonsdale, John. 1981. "States and Social Processes in Africa: A Historiographical Survey." *African Studies Review* 24, nos. 2–3:139–225.

Lussana, Filippo, and Carlo Frua. 1981. "Su la pellagra." In *Pellagra*, edited by Kenneth J. Carpenter, 13–18. Stroudsburg, Pa.: Hutchinson Ross.

MacGaffey, Wyatt. 1981. "African Ideology and Belief: A Survey." *African Studies Review* 24, nos. 2–3:227–74.

MacNeish, Richard S. 1964. "The Origins of New World Civilization." *Scientific American* 211, no. 5:29–37.

Maddalena, Aldo de. 1974. "Rural Europe, 1500–1750." In *The Fontana Economic History of Europe*, vol. 2, *The Sixteenth and Seventeenth Centuries*, edited by Carolo M. Cipolla, 273–353. Glasgow: William Collins.

Major, Ralph H. 1944. "Don Gaspar Casal, François Thiéry, and Pellagra." *Bulletin of the History of Medicine* 16, no. 4 (November): 351–61.

——, ed. 1945. *Classic Descriptions of Disease*. Springfield, Ill.: Charles C. Thomas.

Mancini, Benedetto. 1810. *Del mayz mal coltivato, cagione dello smagrimento della terra, memoria di . . . .* Macerata: Presso Antonio Carlesi.

Mangelsdorf, Paul C. 1974. *Corn: Its Origin, Evolution, and Improvement*. Cambridge: Harvard University Press.

Mangelsdorf, Paul C., Richard S. MacNeish, and Gordon R. Willey. 1954. "Origins of Agriculture in Middle America." In *Handbook of Middle American Indians*, vol. 1, edited by Robert C. West, 427–45. Austin: University of Texas Press.

Manners, Robert A. 1967. "The Kipsigis of Kenya: Culture Change in a 'Model' East African Tribe." In *Contemporary Change in Traditional Societies*, vol. 1, *Introduction to African Tribes*, edited by Julian H. Steward, 205–359. Urbana: University of Illinois Press.

Marie, Armand. 1910. *Pellagra*. Translated by C. H. Lavinder and J. W. Babcock. Columbia, S.C.: The State Company.

Martínez, Maximino. 1959. *Plantas útiles de la flora mexicana*. Mexico: Ediciones Botas.

Masefield, G. B. 1967. "Crops and Livestock." In *The Cambridge Economic History of Europe*, vol. 4, *The Economy of Expanding Europe in the Sixteenth and Seventeenth Centuries*, edited by Edwin E. Rich and C. H. Wilson, 275–301. Cambridge: Cambridge University Press.

Mauro, Frédéric. 1969. *Europa en el siglo XVI: Aspectos económicos*. Barcelona: Editorial Labor.

Mauro, Frédéric, and Geoffrey Parker. 1977a. "Portugal." In *An Introduction to the Sources of European Economic History, 1500–1800*, edited by Charles Wilson and Geoffrey Parker, 63–80. Ithaca, N.Y.: Cornell University Press.

——. 1977b. "Spain." In *An Introduction to the Sources of European Economic History, 1500–1800*, edited by Charles Wilson and Geoffrey Parker, 37–62. Ithaca, N.Y.: Cornell University Press.

McEvedy, Colin, and Richard Jones. 1978. *Atlas of World Population History*. Harmondsworth: Penguin Books.

McNeill, William H. 1976. *Plagues and Peoples*. Garden City, N.Y.: Anchor Press/Doubleday.

Merrill, Elmer D. 1954. "The Botany of Cook's Voyages and Its Unexpected Significance in Relation to Anthropology, Biogeography, and History." *Chronica Botanica* 14, nos. 5–6 (Autumn): 161–384.

Mesa Bernal, Daniel. 1995. *Historia natural del maíz*. Medellín: Vieco y Cía.

Messedaglia, Luigi. 1927. *Il maïs e la vita rurale italiana: Saggio di storia agraria*. Piacenza: Federazione Italiana dei Consorzi Agrari.

Mintz, Sidney W. 1974. *Caribbean Transformations*. Chicago: Aldine.

——. 1985. *Sweetness and Power: The Place of Sugar in Modern History*. New York: Viking.

Miracle, Marvin P. 1966. *Maize in Tropical Africa*. Madison: University of Wisconsin Press.

Mitrany, David. 1930. *The Land and the Peasant in Rumania: The War and Agrarian Reform, 1917–21*. London: Oxford University Press.

Mols, Roger. 1974. "Population in Europe, 1500–1700." In *The Fontana Economic History of Europe*, vol. 2, *The Sixteenth and Seventeenth Centuries*, edited by Carlo M. Cipolla, 15–82. Glasgow: William Collins.

Monardes, Nicolás de. 1921. *Joyful News Out of the New World*. Translated by John Frampton. London: Tudor Translation Series.

Moore, Barrington, Jr. 1966. *Social Origins of Dictatorship and Democracy: Lord and Peasant in the Making of the Modern World*. Boston: Beacon Press.

Morgan, Dan. 1980. *Merchants of Grain*. Markham, Ont.: Penguin Books.

Morgan, Ora S., ed. 1933. *Agricultural Systems of Middle Europe*. New York: Macmillan.

Mundkur, Balaji. 1980. "On Pre-Columbian Maize in India and Elephantine Deities in Mesoamerica." *Current Anthropology* 21, no. 5:676–79.

Murphy, Charles J. 1890. *Lecture on Indian Corn—Maize—as Cheap, Wholesome and Nutritious Human Food, Delivered Before the National Agricultural Society of France*. Edinburgh: R. Grant.

Museo de Culturas Populares. 1982a. *El maíz: Fundamento de la cultura popular mexicana*. Mexico: Museo de Culturas Populares.

——. 1982b. *Recetario mexicano del maíz*. Mexico: Museo de Culturas Populares.

Myrick, Herbert, ed. 1913. *The Book of Corn: A Complete Treatise upon the Culture, Marketing, and Uses of Maize in America and Elsewhere*. New York: Orange Judd.

National Conference on Pellagra. 1910. *Transactions of National Conference on Pellagra, Held under the Auspices of South Carolina State Board of Health at State Hospital for the Insane, Columbia, S.C., November 3 and 4, 1909*. Columbia, S.C.: The State Company.

National Research Council, Committee on Agricultural Efficiency. 1975. *Agricultural Production Efficiency*. Washington, D.C.: National Academy of Sciences.

———. 1981. "Cereal Enrichment in Perspective." In *Pellagra*, edited by Kenneth J. Carpenter, 334–41. Stroudsburg, Pa.: Hutchinson Ross.

Niccoli, Vittorio. 1902. *Saggio storico e bibliografico dell' agricoltura italiana, dalle origini al 1900*. Turin: Unione Tipografico Editrice.

Orleans, Leo A. 1977. *The Role of Science and Technology in China's Population/Food Balance*. Washington, D.C.: Government Printing Office.

Parmentier, Antoine Augustín. 1812. *Le maïs, ou, blé de Turquie, apprécié sous tous ses rapports: Mémoire couronné, le 25 Août 1784, par l'Académie Royale des Sciences, Belles-Lettres et Arts de Bordeaux*. Paris: De l'Imprimerie Impériale.

Parry, John H. 1949. *Europe and a Wider World, 1415–1715*. London: Hutchinson's University Library.

———. 1967. "Transport and Trade Routes." In *The Cambridge Economic History of Europe*, vol. 4, *The Economy of Expanding Europe in the Sixteenth and Seventeenth Centuries*, edited by Edwin E. Rich and C. H. Wilson, 155–219. Cambridge: Cambridge University Press.

Patiño, Victor Manuel. 1963. *Plantas cultivadas y animales domésticos en América equinoccial*. Cali: Imprenta Departamental.

Perkins, Dwight H. 1969. *Agricultural Development in China, 1368–1968*. Chicago: Aldine.

Pickersgill, Barbara, and Charles B. Heiser, Jr. 1977. "Origin and Distribution of Plants Domesticated in the New World Tropics." In *Origins of Agriculture*, edited by Charles A. Reed, 803–35. The Hague: Mouton.

Pimentel, David, and Marcia Pimentel. 1977. "Counting the Kilocalories." *Ceres* 10, no. 5 (September–October): 17–21.

Popol Vuh: Las antiguas historias del Quiché. 1947. Mexico: Fondo de Cultura Económica.

Portères, Roland. 1955. "L'Introduction du maïs en Afrique." *Journal d'Agriculture Tropicale et de Botanique Appliquée* 2, nos. 5–6 (May–June): 221–31.

———. 1959. "Le maïs ou blé des Indes." *Journal d'Agriculture Tropicale et de Botanique Appliquée* 6, nos. 1–3 (January–March): 84–105.

———. 1967. "Iconographíe européenne du maïs." *Journal d'Agriculture Tropicale et de Botanique Appliquée* 14, nos. 10–11 (October, November): 500–501.

Rasmussen, Wayne D., ed. 1960. *Readings in the History of American Agriculture*. Urbana: University of Illinois Press.

Reed, Charles A., ed. 1977. *Origins of Agriculture*. The Hague: Mouton.

Reed, Howard S. 1942. *A Short History of Plant Sciences*. Waltham, Mass.: Chronica Botanica.

Reed, Nelson. 1964. *The Caste War of Yucatan*. Stanford: Stanford University Press.

Riant, Paul Edouard Didier. 1877. *La charte du maïs*. Paris: V. Palme.

Ribeiro, Orlando. 1955. *Portugal*. Vol. 4 of *Geografía de España y Portugal*, edited by Manuel de Terán. Barcelona: Montaner y Simón.

———. 1993. *Portugal, o Mediterrâneo e o Atlântico: Esboções de relações geográficas*. Lisboa: Livraria João Sá da Costa.

Rich, Edwin E. 1967. "Colonial Settlement and Its Labour Problems." In *The Cambridge Economic History of Europe*, vol. 4, *The Economy of Expanding Europe in the Sixteenth and Seventeenth Centuries*, edited by Edwin E. Rich and C. H. Wilson, 302–73. Cambridge: Cambridge University Press.

Richards, Audrey I. 1939. *Land, Labour, and Diet in Northern Rhodesia: An Economic Study of the Bemba Tribe*. London: Oxford University Press.

Riley, Carroll L., et al., eds. 1971. *Man across the Sea: Problems of Pre-Columbian Contacts*. Austin: University of Texas Press.

Robbins, William. 1974. *The American Food Scandal: Why You Can't Eat Well on What You Earn*. New York: Morrow.

Rodney, Walter. 1982. *How Europe Underdeveloped Africa*. Washington, D.C.: Howard University Press.

Rodríguez Vallejo, José. 1976. *Ixcatl, el algodón mexicano*. Mexico: Fondo de Cultura Económica.

Roe, Daphne A. 1973. *A Plague of Corn: The Social History of Pellagra*. Ithaca, N.Y.: Cornell University Press.

Romani, Mario. 1957. *L'Agricoltura in Lombardia dal periodo delle riforme al 1859: Struttura, organizzazione sociale e tecnica*. Milan: Vita e Pensiero.

———. 1963. *Un secolo di vita agricola in Lombardia, 1861–1961*. Milan: Giuffre Editore.

Salaman, Radcliffe N. 1949. *The History and Social Influence of the Potato*. Cambridge: Cambridge University Press.

Salas, Ismael. 1981. "Etiology and Prophylaxis of Pellagra." In *Pellagra*, edited by Kenneth J. Carpenter, 19–24. Stroudsburg, Pa.: Hutchinson Ross.

Salisbury, Neal. 1981. "Squanto: Last of the Patuxets." In *Struggle and Survival in Colonial America*, edited by David G. Sweet and Gary B. Nash, 228–46. Berkeley: University of California Press.

Sambon, L. W. 1981. "Progress Report on the Investigation of Pellagra." In *Pellagra*, edited by Kenneth J. Carpenter, 28–30. Stroudsburg, Pa.: Hutchinson Ross.

Sauer, Carl O. 1962. "Maize into Europe." In *International Congress of Americanists, July 25, 1960*, 777–88. Vienna: Verlags Ferdinand Berger, Horn.

———. 1971. *Sixteenth Century North America: The Land and People as Seen by Europeans*. Berkeley: University of California Press.

Sauer, Jonathan D. 1976. "Changing Perception and Exploitation of New World Plants in Europe, 1492–1800." In *First Images of America*, vol. 2, *The Impact of the New World on the Old*, edited by Fredi Chiappelli et al., 813–32. Berkeley: University of California Press.

Schlebecker, John T. 1975. *Whereby We Thrive: A History of American Farming, 1607–1972*. Ames: Iowa State University Press.

Schneider, Burch H. 1955. "The Nutritive Value of Corn." In *Corn and Corn Improvement*, edited by George F. Sprague, 637–78. New York: Academic Press.

Sereni, Emilio. 1997. *History of the Italian Agricultural Landscape*. Translated by R. Burr Litchfield. Princeton, N.J.: Princeton University Press.

Shannon, Fred A. 1945. *The Farmer's Last Frontier: Agriculture, 1860–1897*. New York: Farrar and Rinehart.

Slicher van Bath, B. H. 1963. *The Agrarian History of Western Europe, A.D. 500–1850*. Translated by Olive Ordish. New York: St. Martin's Press.

Somolinos d'Ardois, Germán, and José Miranda. 1960. *Vida y obra de Francisco Hernández: Precedida de España y Nueva España en la época de Felipe II*. Mexico: Universidad Nacional Autónoma de Mexico.

Sprague, George F., ed. 1955. *Corn and Corn Improvement*. New York: Academic Press.

Stampp, Kenneth M. 1956. *The Peculiar Institution: Slavery in the Ante-Bellum South*. New York: Knopf.

Stavenhagen, Rodolfo. 1975. *Social Classes in Agrarian Societies*. Garden City, N.Y.: Anchor Press/Doubleday.

Steele, Leon. 1978. "The Hybrid Corn Industry in the United States." In *Maize Breeding and Genetics*, edited by David B. Walden, 29–40. New York: Wiley.

Stern, Steve J. 1987. "Feudalism, Capitalism, and the World System in the Perspective of Latin America and the Caribbean." Unpublished manuscript, University of Wisconsin, Madison.

Stoianovich, Traian. 1951. "Le maïs." *Annales* 6, no. 2:190–93.

———. 1953. "Land Tenure and Related Sectors of the Balkan Economy, 1600–1800." *Journal of Economic History* 13, no. 4 (fall): 398–411.

Stonor, C. R., and Edgar Anderson. 1949. "Maize among the Hill People of Assam." *Annals of the Missouri Botanical Garden* 36, no. 3 (September): 355–404.

Stoykovitch, Velimir N. 1933. "The Economic Position and Future of Yugoslavian Agriculture." In *Agricultural Systems of Middle Europe*, edited by Ora S. Morgan, 353–405. New York: Macmillan.

Strauss, Erich. 1969. *Soviet Agriculture in Perspective: A Study of Its Successes and Failures*. New York: Praeger.

Sturtevant, Edward L., and Ulysses Prentiss Hedrick. 1972. *Sturtevant's Edible Plants of the World*. Edited by Ulysses Prentiss Hedrick. New York: Dover Publications.

Sweet, David G., and Gary B. Nash, eds. 1981. *Struggle and Survival in Colonial America*. Berkeley: University of California Press.

Sydenstricker, Edgar. 1915. "The Prevalence of Pellagra: Its Possible Relation to the Rise in the Cost of Food." *Public Health Reports*, October 22, 30, pt. 2:3132–48.

Symons, Leslie. 1972. *Russian Agriculture: A Geographic Survey*. London: G. Bell.

Tawney, Richard H. 1932. *Land and Labor in China*. London: George, Allen and Unwin.

———. 1967. *The Agrarian Problem in the Sixteenth Century*. New York: Harper and Row.

Terris, Milton, ed. 1964. *Goldberger on Pellagra*. Baton Rouge: Louisiana State University Press.

Tracy, Michael. 1964. *Agriculture in Western Europe*. New York: Praeger.

Ucko, Peter J., and G. W. Dimbleby, eds. 1969. *The Domestication and Exploitation of Plants and Animals*. London: Duckworth.

Vance, Rupert. 1929. *Human Factors in Cotton Culture: A Study in the Social Geography of the American South*. Chapel Hill: University of North Carolina Press.

———. 1932. *Human Geography of the South: A Study in Regional Resources and Human Adequacy*. Chapel Hill: University of North Carolina Press.

———. 1937. *Farmers without Land*. New York: Public Affairs Committee.

Vavilov, Nikolai I. 1951. *The Origin, Variation, Immunity, and Breeding of Cultivated Plants: Selected Writings*. Waltham, Mass.: Chronica Botanica Company.

Volin, Lazar. 1970. *A Century of Russian Agriculture: From Alexander II to Khrushchev*. Cambridge: Harvard University Press.

Walden, David B., ed. 1978. *Maize Breeding and Genetics*. New York: John Wiley.

Walden, Howard T. 1966. *Native Inheritance: The Story of Corn in America*. New York: Harper and Row.

Wallace, Henry A., and Earl N. Bressman. 1949. *Corn and Corn Growing*. 5th ed., revised by J. J. Newlin, Edgar Anderson, and Earl N. Bressman. New York: J. Wiley.

Wallace, Henry Agard, and William L. Brown. 1956. *Corn and Its Early Fathers*. East Lansing: Michigan State University Press.

Wallerstein, Immanuel M. 1976. *The Modern World System: Capitalist Agriculture and the Origins of the European World-Economy in the Sixteenth Century*. New York: Academic Press.

———. 1984. *The Politics of the World-Economy: The States, the Movements, and the Civilizations: Essays*. Cambridge: Cambridge University Press.

Warntz, William. 1957. "An Historical Consideration of the Terms 'Corn' and 'Corn Belt' in the United States." *Agricultural History* 31, no. 1:40–45.

Watson, E. J. 1910. "Economic Factors of the Pellagra Problem in South Carolina." In National Conference on Pellagra, *Transactions of National Conference on Pellagra, Held under the Auspices of South Carolina State Board of Health at State Hospital for the Insane, Columbia, S.C., November 3 and 4, 1909*, 25–32. Columbia, S.C.: The State Company.

Weatherwax, Paul. 1923. *The Story of the Maize Plant*. Chicago: University of Chicago Press.

———. 1954. *Indian Corn in Old America*. New York: Macmillan.

Wellhausen, Edwin J. 1978. "Recent Developments in Maize Breeding in the Tropics." In *Maize Breeding and Genetics*, edited by David B. Walden, 59–84. New York: Wiley.

Wellhausen, Edwin J., L. M. Roberts, and E. Hernández X., in collaboration with Paul C. Mangelsdorf. 1952. *Races of Maize in Mexico: Their Origin, Characteristics, and Distribution*. Jamaica Plain, Mass.: Bussey Institution of Harvard University.

Wet, J. M. J. de, and J. R. Harlan. 1978. "Trypsacum and the Origin of Maize." In *Maize Breeding and Genetics*, edited by David B. Walden, 129–41. New York: Wiley.

Will, George Francis, and George E. Hyde. 1964. *Corn among the Indians of the Upper Missouri*. Lincoln: University of Nebraska Press.

Williams, Eric. 1944. *Capitalism and Slavery*. Chapel Hill: University of North Carolina Press.

Withers, Robert S[teele]. 1951. "The Pioneer's First Corn Crop." *Missouri Historical Review* 46, no. 1:39–45.

Wolf, Eric R. 1966. *Peasants*. Englewood Cliffs, N.J.: Prentice-Hall.

———. 1969. *Peasant Wars of the Twentieth Century*. New York: Harper and Row.

———. 1982. *Europe and the People without History*. Berkeley: University of California Press.

Yen, D. E. 1959. "The Use of Maize by the New Zealand Maoris." *Economic Botany* 13, no. 4:319–27.

Zeichner, Oscar. 1939. "The Transition from Slave to Free Agricultural Labor in the Southern States." *Agricultural History* 13, no. 1:22–32.

INDEX

Abolitionists, 63

Achras zapota Linn., 3

Acosta, José de, 24, 98

Adams, Thomas, 3–4

Afghanistan, 41

Africa: sisal in, 10, 63; corn in diets in, 24, 60, 65, 66, 70, 71–72, 83, 88–89, 95; names for corn in, 30, 60, 61; and slave trade, 51–60, 62–65, 66, 67, 175, 237–38; and foreign trade, 52, 67–68, 77–78, 83–87; firearms imported to, 56; population growth and demographic development in, 58–59, 82–83, 85; economic patterns in, 59; corn's introduction and diffusion in, 60–62, 66–67; corn cultivation and production in, 60–64, 66–67, 68, 71–73, 81, 87–96, 118; corn yields in, 63–64, 72, 73, 77, 89–90, 92–93; agriculture in, 63–67, 70, 72–77, 79–81, 83–88, 90–92; rice cultivation in, 64, 71, 92, 93; colonialism in, 66–81, 83–84; secondary states in, 68; Zulu empire in, 68; Boers in, 68, 69; Creoles in, 68, 69; diseases in, 69, 73; food rations in, 70–71; native workforce in, 70–71, 73–74, 79–80; discriminatory regulations in, 70–71, 74–75; and warfare, 71–72; famines in, 72; mines in, 72; white agriculturalists in, 72–76, 90–91; stock breeding in, 73, 76; corn market in, 74–75, 90, 95; plantations in, 75–76, 86; railroads in, 76, 78–79; enclosure movement in, 76–77; taxation in, 77–78; transporting corn in, 78–79; money economy in, 79, 80, 81, 95; land tenure systems in, 80; government and eco-nomic models of modern states in, 82; diets in, 82–87, 160–61, 204; independence from colonial rule in, 82–96; corn exports from, 83, 84, 88; food exports from, 83, 85–86, 87; food imports to, 84, 85, 86–87, 213, 214; per capita food production in tropical Africa, 85; violence during 1970s in, 87; pellagra in, 138, 142, 150. *See also specific countries*

Afrikaaners, 69

Agave spp., 8, 10

Agrarian reform, 227, 231

Agricultural Trade Development and Assistance Act (1954), 86, 190–91, 205–6

Agriculture: value of American crops, 1, 190; top seven crops, 1–2, 6; development of, and domestication of plants, 4–5; statistics on corn production, 12, 44, 87–90, 92–93, 97, 159, 180–81, 183, 187, 188–89, 192, 202, 212, 217, 218, 225, 230; yields for corn, 15, 16–17, 19, 44, 63–64, 72, 73, 77, 89–90, 92–93, 109, 117–18, 180–81, 183, 184, 185, 187, 192, 200, 202, 218–19, 221, 227, 229–30; yields for wheat, 15, 183, 192, 199, 221, 227, 230; calculation of agricultural yields, 15–16; yields for cereals, 15–16, 63–64, 117–18, 183, 192, 199, 221, 227; and hybrid seed, 19, 184–85, 187, 201, 235; slash-and-burn agriculture, 41, 64, 76; in China, 42–50, 213; and share-cropping, 47, 127, 147–48, 164–65, 166–68, 173, 184, 238; in Africa, 63–65, 70, 72–77, 79–81, 83–88, 90–92; and insect infestations, 72, 76, 138; plow and draft

animals for, 73, 154, 180, 184; and soil erosion, 77; and mechanization, 91, 180, 194, 219; in Europe, 97, 102–3, 112–31, 147, 149, 221; and agronomy texts in Europe, 102–3; in Spain, 104–6, 115; and irrigation, 105–6, 113–14, 115, 187, 194; in Portugal, 106–7, 115; in France, 107, 109, 115–16, 139; in Italy, 107–8, 125, 146–47; in Balkans, 108–9; in Russia, 109; and plow, 113, 127, 180; and crop rotation, 114, 115–16, 118, 135, 221; and fertilizers, 115, 166, 185–86, 187, 194, 201, 220–22, 225–26, 228, 229, 230; three-field system of, 116–17; scientific research on, 119, 185, 186, 193, 195, 201, 228–31; in English colonies in America, 151–57, 175–76; and Native Americans, 153–54; in United States, 156–67, 172–75, 178–96, 212, 218–22; furnishing system for, 165–66, 167–68, 238; and farm machinery, 180, 184, 187–88, 219, 229; and tractors, 184; government programs and subsidies for, 186–87, 190–91, 195, 210, 211, 222; and genetic engineering, 187, 219–20, 228; and herbicides, 187, 221; scientific or capitalized intensive agriculture for corn production, 187–88, 217–24, 235; profitability of, in United States, 195–96; in Soviet Union, 198–203, 211; future trends in corn production, 217–31; traditional peasant agriculture for corn production, 218, 224–31; and Green Revolution, 223–24, 227, 235; and insecticides, 228. *See also* Corn; *and specific crops*
Agronomy, 102–3, 134, 185, 201
Alabama, 161, 168
Albania, 108, 140
Alcoholic drinks, 2, 22–23, 42, 77, 155, 161, 175, 176–77
Alegria, 7
Alembert, Jean Le Rond d', 103
Algeria, 138
Allende, Salvador, 207
Alliance for Progress, 227

Amaranthus cruentus Linn., 7
America. *See* American plants; New World; United States
American plants: top seven plants, 1–2, 6; uses of, 2–4; domestication of, 4–5; and diets of pre-Hispanic population, 5–8; description of, 6–11; maintenance foods, 6, 9; as luxuries, 8–9; fruits, 9, 154; textile plants, 10–11; origin of corn in America, 28–36, 98–99; corn discovered by Columbus, 37; in Africa, 63–64, 83; in herbals, 99–102; and English colonists in America, 151–57. *See also* Corn; *and other specific plants*
Ananas comosus Linn., 9
Anderson, Edgar, 40
André company, 207–10
Andropogon sorghum Brotero, 60
Anglo-Franco War (1860–61), 48
Angola, 58, 61, 73, 88
Animals: domesticated animals, 5; corn as feed for, 22, 24–25, 26, 89, 97, 105, 110, 130, 155, 176, 181, 188–89, 235; livestock in Africa, 73, 76; plow and draft animals, 73, 154, 179, 180, 184; livestock in Europe, 109, 110, 113, 115, 116, 128, 130; dairy animals, 154; in English colonies in America, 154; livestock in United States, 154, 176, 189–90, 194; livestock in Third World countries, 226. *See also* Meat in diets
Annatto, 8–9
Anthropology, vii, xii
Antilles, 54, 62, 114
Arachis hypogaea Linn., 9
Archaeological research, 31–33
Argentina, 202, 203, 207, 209, 210
Arkansas, 168
Asante, 62–63
Assam, 40–41
Australia, 203, 221
Austria, 97, 109, 138, 141
Avocado, 9, 32
Aztec creation myth, 35

Babcock, James, 137

Balkans, 97, 108–9, 110, 116, 129, 130, 131

Bangladesh, 40

Bantu groups, 68

Barbasco, 2

Barley, 1, 97, 106, 115, 116, 118, 136, 191, 212, 213

Bastard identity of corn, xiii, 28–37, 241

Beadle, George, 34

Beans, 154. *See also* Frijoles

Belgium, 69, 79

Benin, 88

Bible, 31

Birth control pill, 2

Bixa orellana Linn., 8–9

Bock, Jerome, 100

Boers, 68, 69

Bonafous, Matthieu, 38, 103

Bonfil, Guillermo, x

Born, Marcos and Leonora, 209

Boserup, Ester, 122

Bosnia, 108, 138

Boxer Rebellion, 48

Bractea, 21

Braudel, Fernand, 104

Brazil: fuel in, 23; and Portuguese trade route, 41; and slavery, 54, 61–62, 67, 159; per capita income in, 122; and world agriculture market, 207, 209, 210, 212; poor peasants in, 215

Brazilwood, 2

Britain. *See* England

Brooks, Eugene, 179

Bucovina, 138

Buda, Gaetano Harasti da, 103

Buffon, Georges-Louis, 102

Bulgaria, 97, 108, 138

Bunge Corporation, 207–10

Buniva, Professor, 134

Burma, 39, 40

Bursera jorullensis Engler, 8

Burundi, 79

Cabeza de negro, 2

Cacao, 8, 63, 75, 79, 80, 85, 86

Calocarpum mammosum Linn., 9

Caloric value of corn, 92–93, 189

Cambodia, 206, 207

Cameroon, 88

Canada, 203, 212

Canals, 179

Capitalism, 128, 149–50, 207–10, 233–34, 237–40

Capitalized intensive or scientific agriculture, 187–88, 217–24, 235

Capsicum annuum Linn., 7

Cárdenas, Juan de, 12–13, 19–24, 26

Cargill Inc., 207–10

Caribbean, 54, 158. *See also specific countries*

Carica papaya Linn., 9

Carrera Pujal, Jaime, 105

Casal, Gaspar, 132–33

Cassava, 1–2, 6, 63, 64–65, 79, 83, 88, 93, 94

Catkin, 3

Cereals. *See specific crops*

Cheng Han-seng, 47

Chenopodium quinoa Willdenow, 7

Chestnuts, 110

Chewing gum, 3–4

Chicha, 22

Chile, 207

Chili peppers, 7, 32, 42

China: diets in, 24, 42; food needs in, 25; references to corn in, 31; woodcut of corn from, 38–39; earliest references to corn in, 39; corn's introduction and diffusion in, 39–42; names for corn in, 40; food plants in, 42; sweet potatoes in, 42; rice production in, 42, 43, 44, 46, 50; corn cultivation and production in, 42–44, 118, 213; agriculture in, 42–50, 213; population of, 43, 44–45; wheat production in, 44; second agricultural revolution in, 44–46, 48, 50; famines and hunger in, 45; land division and tenure in, 45–47; inheritance in, 46–47; sharecropping in, 47; trade between Europe and, 48; violence in, during nineteenth century, 48–49; in early twentieth century, 49; silver in, 49; revolution of 1949 in, 50; corn imported to, 214

Chocolate, 8

Cieza de León, Pedro de, 98

Cinchona spp., 2

Civil War (U.S.), 163

Coca, 9–10

Cocaine, 10

Cochineal, 2, 5, 8

Cocoa, 9

Coffee, 75, 79, 80, 85, 86

Colgate & Company, 182

Colombia, 204–5

Colonial America, 151–58, 174–76

Colonialism: and Mexicanism, xi; in Africa, 66–81, 83–84. *See also* Slavery and slave trade

Columbus, Christopher, 8, 30, 37

Congo, 58, 61, 69, 70, 73, 74

Constantinople, 108

Continental Grain Company, 207–10

Contraceptives, 2

Copal, 8

Corfu, 140

Corn, 232–41; bastard identity of, xiii, 28–36, 241; scientific nomenclature for, 12; universal appeal of, 12–13, 15, 19–27, 38, 44, 73, 77, 93–95; botanical economy of, 12–27; general cultivation of, 13–14, 15, 17–19, 26–27, 34–35; physical characteristics of, 14, 16–17, 26–27, 185; varieties of, 14–15; kernels of, 16–17, 22, 27, 64; picking of, 17; ear of, 17, 27; roots of, 17, 114–15; distance between corn plants, 17–19; sowing of, 17–19; hybrid seed for, 19, 184–85, 187, 235; uses of, 21–27, 155, 188–89, 235–36; as animal feed, 22, 24–25, 26, 89, 97, 105, 110, 130, 155, 176, 181, 188–89, 235; refining industry for, 25–26, 236; no wild varieties of, 26–27; origins of, 28–36, 60, 98–99, 103, 104; names for, 29–31, 40, 61; migration of, in fifteenth and sixteenth

centuries, 37–38; and slave trade, 62–64, 160–61; caloric value of, 92–93, 189; New World chroniclers of, 98–100; prices for, 106, 107–8; diverse processes associated with, 233–40. *See also* Agriculture; Diets; *and specific countries and continents*

Corn bread, 154–55

Corncobs, 21–22, 155

Corn flour, 71, 95, 109–10, 131, 134, 145, 182, 189

Corn husks, 21, 27, 155

"Corn Letter," 31

Corn liquor, 23, 42, 155, 161, 176–77

Corn mash, 155

Cornmeal, 155

Corn mush, 155, 160–61

Corn oil, 26, 182

Corn paste, 89, 110, 131

Corn silks, 22

Corn smut, 22, 219

Cornstalk, 21, 115, 130

Cornstarch, 182

Corn sugar, 26

Cortisone, 2

Cotton: as leading crop, 9; Egypt's production of, 10, 142; varieties of, 10–11; and textile industry, 11, 159, 170–71; Africa's production of, 85; Europe's production of, 114; U.S. production of, 157, 159, 161, 163, 164, 166–69, 171–75, 206; imports of, to England, 159

Couscous, 160

Creation myths, 35

Creoles, 68, 69

Croatia, 108, 138

Crops. *See* Agriculture; Corn; *and other specific crops*

Crosby, Alfred, 122

Cuba, 37, 54, 67

Cucurbita spp., 7, 18

Cuitlacoche, 22

Curtin, Philip D., 53, 57

Cyprus, 52

Dahomey, 68
Dalmatia, 108, 138
De Candolle, Alphonse, 28–29
Deere, John, 180
De Herrera, Gabriel Alonso, 103
Demography. *See* Population
De Monardes, Nicolás, 2
Denmark, 54, 55
De Pauw, Cornelius, 102
Developing countries. *See* Africa; Latin
  America; World market; *and specific
  countries*
Dextrines, 26
Dextrose, 26
Diderot, Denis, 103
Diets: of pre-Hispanic population, 5–8;
  corn in, 6, 19–24, 25, 42, 60, 65, 66, 70,
  71–72, 83, 88–89, 95, 105, 107, 109–11,
  122–23, 131, 132, 134–37, 143–45, 148,
  154–55, 159–61, 167, 176, 189, 214, 236; in
  China, 24, 42; meat in, 25, 144, 154, 155,
  160, 161, 167, 176, 189–90, 194, 198–99,
  213, 222; industrialized foods, 25–26;
  and supermarkets, 26; in Africa, 82–89,
  160–61, 204; in Europe, 105, 107, 109–11,
  121–23, 131, 133–37, 143–45, 148; poverty
  associated with corn in, 110–11, 123–24,
  131, 133; and pellagra, 133–37; in Mexico,
  139, 143, 177, 236; dairy products in, 144,
  189; fats in, 144; vegetables in, 144, 189–
  90; in English colonies in America,
  154–55; in United States, 159–62, 167–
  68, 170, 172, 176, 189–90; of slaves, 160–
  61; protein in, 189–90, 198–99, 211; in
  Soviet Union, 198–99; in Dominican
  Republic, 204
Dioscorea composita Hemsley, 2
Dioscorea mexicana Guillemin, 2
Diseases, 52, 69, 73, 103, 111, 119, 121, 132–
  50, 152, 168–73. *See also* Pellagra
Dolado, Juan Bautista, 133
Domesticated animals. *See* Animals
Dominican Republic, 37, 204
Dyes, 2, 8

Ear of corn, 17, 27
East Africa, 58, 68
Eastern Europe, 202, 203
Ecuador, 207
Egypt, 30, 68, 113, 142
Elvehjem, Conrad A., 169
Energy resources, 22, 23, 194–95
Engerman, Stanley, 161
England: wheat production in, 16; corn
  cultivation in, 22; name for corn in, 30;
  and China, 48; and slave trade, 54, 55,
  66, 67, 158; and colonialism in Africa,
  69, 79; and corn's origins, 101; popula-
  tion of, 120; 1750 per capita income in,
  122; colonies of, in America, 151–57,
  174–76; cotton imported to, 159; corn
  imported to, 182; agricultural exports
  from, 212
Erythroxylum coca Lamarck, 9
Ethiopia, 69
Ethyl alcohol, 23
Euchlaena mexicana, 29, 34
Euphorbia cerifera, 3
Europe: and Industrial Revolution, 6, 10,
  11, 54, 121, 125, 159, 238; wheat production in, 15, 16, 97, 115, 116–18, 213, 214;
  livestock in, 15, 109, 110, 113, 116, 128,
  130; cereal production in, 15–16, 117–18,
  121, 129–30, 135; and trade, 48, 52, 67–
  68, 125, 130; and slave trade, 51–60, 62–
  64, 66, 67, 175, 237–38; agriculture in,
  97, 102–4, 112–31, 147, 149, 221; corn
  cultivation and production in, 97, 112–
  18, 129–32, 134–35, 147–50; corn's introduction and diffusion in, 97–113, 132;
  and New World chroniclers, 98–100;
  and herbals, 99–102; agronomy treatises in, 102–3; pellagra in, 103, 111, 121,
  132–50, 160; diets in, 105, 107, 109–11,
  121–23, 131, 133–37, 143–45, 148; corn in
  diets in, 105, 107, 109–11, 122–23, 131,
  132, 134–37, 143–45, 148; irrigation in,
  105–6, 113–14, 115; poverty in, 110–11,
  123–24, 131, 133, 146–50; revolution in,
  111, 128; factors in corn's diffusion in,

Khrushchev, Nikita, 198–202
Kingsford, Thomas, 182
Kipsigis, 76, 80
Komenda, 62

Lalesque, 143
Landownership: in China, 45–47; and African plantations, 75–76, 86; in Africa, 80; in Europe, 126–29, 146–49; in Italy, 146–48; and pellagra, 146–49; in France, 148; in colonial America, 156–57; and U.S. plantations, 157–64, 174–76; in United States, 157–64, 174–76, 178, 184
Lane, John, 180
Langer, William, 122
Larrea divaricata Cavanilles, 3
Las Casas, Friar Bartolomé de, 53, 98
Latin America, 24, 197, 202, 204–5, 211, 214, 227. *See also specific countries*
Lavinder, Claude H., 137, 169–70
Laxatives, 2
Leach, Edward R., 41
League of Nations, 67, 70
Leaves of corn plant, 22, 155
Leopold II, King, 69
Lever (W. H.) company, 75
Liberia, 67, 68, 69, 75
Lincoln, Abraham, 153
Linnaeus, Carolus, 12, 102
Linseed, 114
Li Shi-Chen, 38–39
Livestock. *See* Animals
Lombroso, Cesare, 136, 137
Lopes, Duarte, 61
López de Gómara, Francisco, 98
Lophophora williamsii Lemaire, 3
Louis-Dreyfus company, 207–10
Louisiana, 168
Lumber, 174
Lussana, Filippo, 135–36
Lycopersicon esculentum Miller, 7
Lysenko, Trofim, 201

Maares, P. de, 61

Madagascar, 73
Magueys, 8
Maize. *See* Corn
Malaria, 2, 69
Malawi, 79, 88
Malnutrition. *See* Famines; Hunger
Malthusian dilemma, 85, 121, 210
Maltodextrines, 26
Mammees, 9
Mangelsdorf, Paul, 32, 33, 34
Manihot esculenta Crantz, 6
Manioc, 63
Maple syrup and maple sugar, 154
Marie, Armand, 137
Martinique, 54
Mártir de Anglería, Pedro, 37
Martyr, Peter, 98
Marzari, Giovanni Battista, 133–34, 143
Massachusetts, 152, 156
Matthiolus, Petrus, 100, 107
Mayas, 10, 35
Maydeae, 12, 34
McCormick, Cyrus, 180
McEvedy, Colin, 58
McNeish, Richard, 31–32
Meat in diets, 25, 144, 154, 155, 160, 161, 167, 176, 189–90, 194, 198–99, 213, 222
Mechoacan root, 2
Medicinal uses of plants, 2, 4, 22
Mediterranean area, 113–15, 123, 125, 130, 138
Mendel, Gregor, 201
Méndez de Cancio, Gonzalo, 105, 132
Messedaglia, Luigi, 31, 100, 103–4
Mexicanism, x–xi
Mexico: corn production and consumption in, ix–x, 15, 18, 21–24, 31–34, 39, 136, 138–39, 143, 145, 177, 181, 236; and Mexicanism, x–xi; plants from, 2, 3, 7, 10, 18; cochineal from, 8; archaeological research in, 31–33; teosinte from, 33–34, 219; *peso* of, 49; Hernandez in, 98–99; 1960s per capita income in, 122; pellagra in, 138–39, 145; diseases in, 138–39, 145, 168, 169; diets in, 139, 143, 177,

in urban areas, 142, 169–70; and peasants, 142–43, 146–50; and poverty, 142–43, 146–50, 170–71; and niacin deficiency, 144, 145, 160, 169; and women, 144, 170; treatment for, 145, 150, 169; and landownership, 146–49; and modernization, 150; and slaves in United States, 160

Peloponnesus, 108

Penicillin, 2

Pennisetum, 60

Persea americana Mill., 9

Peru, 17, 39

Peyote, 3

Phaseolus spp., 7, 18

Pig-corn cycle, 155, 161, 176, 179

Pigweed, 18

Pilgrims, 152–53

Pimentel, David and Marcia, 188, 194

Pineapple, 9

Pinochet, Augusto, 207

Pioneers and frontier, 155–57, 177–80

Plague, 119, 121

Plantations: in Africa, 75–76, 86; in southern United States, 157–64, 174–76

Plants. *See* American plants; Corn; Rice; Tobacco; Wheat; *and other specific plants*

Plow, 113, 127, 180

Plymouth colony, 152–53

Pocahontas, 152, 157

Poland, 138

Polenta, 134, 155

Polo, Marco, 42

Pone, 154

Popol Vuh, 35

Population: of China, 43, 44–45; of slaves, 51–54, 158–59; of Africa, 58–59, 82–83, 85; of Europe, 119–22, 125, 141, 146, 149; of England, 120; of Italy, 120, 141; of colonial America, 156; of United States, 156, 158–59, 183–84, 188; overpopulation, 210; of Latin America, 214

Portères, Roland, 100

Portugal: name for corn in, 30, 60; and

corn's introduction into China, 39, 41; and slave trade, 52, 54–55, 64, 66; and corn's introduction into Africa, 60–62; and colonialism in Africa, 69, 82; corn cultivation and production in, 97, 106–7, 115, 214; corn as decorative element in architecture in, 100–101; writings on corn from, 104; pellagra in, 138; wheat production in, 214

Potatoes, 1–2, 6, 79, 102, 111, 122, 123, 189

Poverty: corn as food associated with, 110–11, 123–24, 131, 133; in Europe, 110–11, 123–24, 131, 133, 146–50; and pellagra, 142–43, 146–50, 170–71; in United States, 170–71, 193–94, 195; in Latin America, 211; in Third World countries, 211, 230; and failure to plan for future, 230; and capitalism, 238, 240

Powhatans, 151, 152

Prejudice, 40–41

Price revolution, 123–24

Prices: of corn, 106, 107–8; of wheat, 106, 108, 124, 130, 197, 213; of oil, 197

Psidium guajava Linn., 9

Puerto Rico, 208

Quetzalcóatl, 35

Quinoa, 7

Railroads, 76, 78–79, 179

Ramusio, Giovanni Battista, 61, 100

Reagan, Ronald, 195

Reconstruction, 163–64, 166–67

Refining industry for corn, 25–26, 236

Ribeiro, Orlando, 100

Rice: importance of, as food, 1; yields for, 16, 227, 230; China's production of, 42, 43, 44, 46, 50; Africa's production of, 64, 71, 92, 93; compared with corn flour, 95; Europe's production of, 114; nutrition of, 136; U.S. production of, 157, 159, 174–75; U.S. exports of, 191, 205, 212; international market for, 205, 212, 213; Third World countries' production of, 229

Robots, 219
Rolfe, John, 152, 157
Romania, 97, 108, 117, 138, 141–42, 148
Roosevelt, Franklin D., 173, 185, 186, 199
Roots of corn plant, 17, 114–15
Roussel, Théophile, 136, 139
Rubber, 2–3, 63, 75
Rum, 175, 176, 177
Russia. *See* Soviet Union
Russo-Japanese War, 48
Rwanda, 79
Rye, 106, 115, 116–17, 118, 136

Saccharum officinarum Linn., 54
Sahel, 85
Sahugún, Friar Bernardino de, 99
Salas, Ismael, 136–37, 138
Sambon, Louis, 142
Sandwith, Fleming, 142
Sanzio, Rafael, 100
Sapodillas, 3, 9, 32
Saudi Arabia, 211
Scandinavia, 22, 54, 55, 120
Scientific or capitalized intensive agriculture, 187–88, 217–24, 235
Scientific research on agriculture, 119, 185, 186, 193, 195, 201, 228–31
Searcy, George H., 168
Sea transport, 179
Seed companies, 185
Senegal, 68
Serbia, 138
Sesame, 7, 32
Setaria italica Beauvois, 60
Sharecropping, 47, 127, 147–48, 164–65, 166–68, 173, 184, 238
Sicily, 52, 126
Sierra Leone, 68
Silk, 157
Silver, 49
Simmondsia chinensis Link, 3
Sisal, 10, 63
Slavery and slave trade, 51–60, 62–64, 66, 67, 157–63, 175, 237–38
Slicher van Bath, B. H., 117

Smallpox, 152
Solanum tuberosum Linn., 6
Sorghum, 30, 64, 71, 92–93, 109, 110, 118, 205
Soursops, 9
South Africa, 69, 70, 73, 75, 76, 82, 83, 87–88, 138
South Carolina, 157, 161, 168, 170
South Dakota, 181
Soviet Union, 87, 97, 109, 138, 197–203, 206, 207, 211, 213, 214
Soy beans, 192, 210, 212
Spain: name for corn in, 30; and slave trade, 52, 53, 55, 67; corn cultivation and production in, 101, 104–6, 110, 115, 132; agronomy book from, 103; land reforms in, 128; corn in diets in, 131, 132; pellagra in, 132–33, 138, 139–40; Squanto in, 152; multinational cereal company in, 207, 209; corn imported to, 214
Squanto, 151, 152–53
Squash, 7, 18, 32, 154
Stalin, Joseph, 198, 201
Stalk of corn, 21, 115, 130
Starch, 26
Steamboats, 179
Stoianovich, Traian, 117
Stonor, C. R., 40
Strambio, Gaetano, 140
Sudan, 58, 68, 88
Sugar, 75, 121, 189
Sugar cane, 23, 54, 85, 111, 114, 138, 157, 159
Sunflowers, 9, 154
Supermarkets, 26
Sweden, 54, 55
Sweet potatoes, 1–2, 6, 42, 79, 88, 93
Swine-corn cycle, 155, 161, 176, 179
Switzerland, 207
Sydenstricker, Edgar, 169–71
Syllacio, Nicolò, 98
Syrups, 26, 154. *See also* Molasses

Taiping Rebellion, 48
Tanner, William F., 169
Tanzania, 88, 91, 95

culture in, 186; government programs for agriculture in, 186–87, 190–91, 195, 210; subsidies for farmers in, 187, 190, 195, 210, 222; contradictions in agriculture in, 193–96; profitability of agriculture in, 195–96; and food aid to poor countries, 204–8; multinational cereal companies in, 207–10; crop land reserve program in, 221

U.S. Department of Agriculture (USDA), 186, 211

U.S. Public Law 480, 86, 190–91, 205

Upper Volta, 88

USDA. *See* U.S. Department of Agriculture

USSR. *See* Soviet Union

Valley of Tehuacán, Puebla, Mexico, 32–33, 35

Vanilla, 8

Vavilov, Nikolai I., 5, 29

Venezuela, 208

Vietnam, 48, 206, 207

Virginia, 151–52, 156, 157

Wallace, Henry, 185, 186, 199

Watson, E. J., 168–69

Wax, 3

Weatherwax, Paul, 33

We Come to Object (Warman), vii

Weismann, August, 201

West Africa, 52, 58, 60, 61, 64, 66–69, 72, 79, 80, 92

Western Europe. *See* Europe

West Indies, 174, 175

Wheat: importance of, as food crop, 1; Europe's production of, 15, 16, 97, 115, 116–17, 118, 213, 214; yields of, 15, 183, 192, 199, 221, 227, 230; England's production of, 16; China's production of, 44; compared with corn in Africa, 71; imported to Africa, 84, 86; prices of, 106, 108, 124, 130, 197, 213; nutrition of, 136; U.S. production and exports of, 183, 191, 192, 205–6, 212; caloric value

of, 189; in world market, 197, 202–3, 205–6; Soviet Union's production of, 199–200; Saudi Arabia's exports of, 211; Europe's exports of, 213; Mexico's production and exports of, 223–24, 227, 228; Third World countries' production of, 229

Whiskey, 23, 42, 155, 161, 176–77

W. H. Lever company, 75

Williams, Eric, 56–57

Wolf, Eric R., xii, 49, 57

World Bank, 211

World hunger. *See* Hunger

World market: and corn, 23, 83, 84, 88, 181–82, 190–93, 202, 203, 205, 211–14; and U.S. exports, 23, 86, 159, 181–83, 190–93, 202, 203, 211, 212, 221–22; grain sales to Soviet Union, 87, 197–98, 201–2, 206; and wheat, 183, 191, 192, 202–3, 205–6, 211–13, 223–24; and oil prices, 197; and wheat prices, 197; and Soviet agriculture, 198–203, 211, 213; grain market before World War II, 202–3; grain market after World War II, 203–4; and U.S. food aid to poor countries, 204–8; and rice, 205, 212, 213; and multinational cereal companies, 207–10; and controls on food exports, 210; and food reserves in exporting countries, 210–11; grain market after 1972 crisis, 210–16, 221–22; and barley, 212, 213; and changes in U.S. foreign trade, 212–13; and food dependence in poor countries, 214–16, 222–23; and hunger, 215–16. *See also* Trade

World War I, 69, 72

World War II, 23, 69–70, 72, 142, 172, 190, 198

Xalapa root, 2

Ximenez, Francisco, 99

Yams, 88, 93

Yellow fever, 69, 169

Yields: for corn, 15, 16–17, 19, 44, 63–64,